Springer Series in Microbiology

Editor: Mortimer P. Starr

Giancarlo Lancini
Francesco Parenti

Antibiotics

An Integrated View

With 106 Figures

Springer-Verlag
New York   Heidelberg   Berlin

Giancarlo Lancini
Director, Preclinical Research
Gruppo Lepetit
Laboratori di Ricerca
20158 Milan
Italy

Francesco Parenti
Director, Antibiotics Research
Gruppo Lepetit
Laboratori di Ricerca
20158 Milan
Italy

Series Editor:
Mortimer P. Starr
Department of Bacteriology
University of California, Davis
Davis, California 95616, U.S.A.

Production: William J. Gabello

The figure on the front cover is a schematic representation of the interaction between valinomycin and K^+.

Library of Congress Cataloging in Publication Data
Lancini, Giancarlo.
    Antibiotics, an integrated view.
    (Springer series in microbiology)
    Bibliography: p.
    Includes index.
    1. Antibiotics. I. Parenti, Francesco. II. Title.
III. Series. [DNLM: 1. Antibiotics. QV 350 L249a]
RM267.L34   615'.329   82-3384
Translated by Betty Rubin, M.D.
This is a revised and expanded translation of the original Italian edition,
Biochimica e Biologia degli Antibiotici, © 1977 by ISEDI, Instituto
Editoriale Internazionale, Milan, Italy.

© 1982 by Springer-Verlag New York, Inc.
Softcover reprint of the hardcover 1st edition 1982

9 8 7 6 5 4 3 2 1

ISBN-13: 978-1-4612-5676-2   e-ISBN-13: 978-1-4612-5674-8
DOI: 10.1007/978-1-4612-5674-8

To Piero Sensi
   discoverer of rifamycins,
   teacher, and friend

Preface

Antibiotics are among the most widely prescribed drugs in both human and veterinary medicine. Furthermore, they are used to protect plants against bacterial and fungal diseases, to decontaminate the shells of eggs, and to improve weight gain and feed conversion in a variety of food animals. Many antibiotics, in addition, have been essential tools in the elucidation of specific cellular functions. Genetic engineering, for example, would not be what it is today without the use of antibiotics in the selection of easily determined genetic markers.

Production of antibiotics involves a diverse group of professionals: the fermentation technologist, the bioengineer, the extraction chemist. To improve productivity, an understanding of the biosynthetic pathway and the mechanisms of its control is often useful.

After the more than 40 years since the discovery of penicillin, the biologist is still unable to answer basic questions: Why are antibiotics produced by only a small number of microbial groups? What is the function of antibiotics in nature?

When we started to teach our course on the science of antibiotics at the University of Pavia and the University of Milan, we realized that there was no book that presented the basic facts and concepts on *all* aspects of this diverse science. This book therefore arose out of our teaching need. Our experience in the discovery, development, and production of antibiotics has certainly imparted a practical nuance to this book.

Each chapter has a list of basic references. However, we have chosen not to cite the names of scientists in the text: doing so would have resulted in too many omissions.

Many colleagues have read and commented on the first draft; their suggestions have been incorporated in the final version, particularly those of

Richard Elander, Richard White, Arnold Demain, and the editor of this series, M. P. Starr. S. Nani supplied literature and a basic scheme for the chapter on veterinary use. The chapter on medical use was reviewed by E. Robotti. B. Rubin revised and corrected the English version. G. Bergonzi, M. P. Biandrate, J. Cole, and C. Holliday patiently typed the many versions of the manuscript.

The comments and suggestions of the many scientists who read the draft, together with those of our students who have used the Italian version, have helped, we hope, to make this book an integrated analysis of the diverse science of antibiotics. To all of them go our warmest thanks.

Pavia, 1982   G. C. LANCINI
Milan, 1982   F. PARENTI

Contents

Chapter 1

The Antibiotics: An Overview

I. Definition

Antibiotics are low molecular weight microbial metabolites that at low concentrations inhibit the growth of other microorganisms. The term low molecular weight substances refers to molecules of at most a few thousand daltons. We do not include among the antibiotics those enzymes, such as lysozyme, and other complex protein molecules that also have antibacterial properties. If the definition were to be adhered to rigorously, the only substances to be considered antibiotics would be the natural products of microorganisms. However, we now include in the category the following semisynthetic antibiotic substances:

1. products obtained by chemical modification of natural antibiotics or of other products of microbial metabolism;
2. products obtained by microbiological transformation of synthetic compounds.

Even the requirement that antibiotics be microbial products is no longer strictly applied. For example, one often sees expressions such as "antibiotic products of plants," but we feel that this usage of the term is inaccurate.

When we speak of inhibition of the growth of other microorganisms by an antibiotic, we mean either temporary or permanent inhibition of the capacity of the microorganism to reproduce and, consequently, inhibition of the growth of the bacterial population rather than of an individual cell. When the inhibition is permanent, the antibiotic activity is termed *bac-*

tericidal or, often, *cidal*. If the inhibition is lost when the antibiotic is removed from the medium, the antibiotic is said to have a *bacterostatic* or *static action*.

The limiting phrase "at low concentration" is added to the definition because, obviously, even normal cell components can cause damage at excessive concentrations. For example, glycine, one of the constituents of every protein, has a strong bactericidal effect on some bacteria, if it is present in the culture medium in concentrations of the order of 3%. Similarly, we do not consider ethyl alcohol, the fermentation product of some microorganisms, to be an antibiotic, because it demonstrates its antibacterial activity only at high concentrations. When we speak of "low concentrations," generally we mean values of < 1 mg/ml.

II. Chemical Nature

To date, no fewer than 3,000 antibiotics have been isolated and described, and the chemical structures of many of them have been determined. With regard to the others, activities and physicochemical properties are known that serve to identify the molecule. Sufficient information is available to indicate that the antibiotics are a very heterogeneous group chemically, in fact:

1. The group includes substances of molecular weights from 150 to 5000 daltons;
2. the molecules may contain only carbon and hydrogen, or, more commonly, carbon, oxygen, hydrogen, and nitrogen; others also contain sulfur, phosphorus, or halogen atoms;
3. almost all the organic chemical functional groups are represented (hydroxyl, carboxyl, carbonyl, nitrogen functions, etc.), as are all the organic structures (aliphatic chains, alicyclic chains, aromatic rings, heterocycles, carbohydrates, polypeptides, etc.).

The only property that all antibiotics have in common is that they are organic solids, which is obvious from their definition as products of microbial metabolism. It is less obvious why liquid antibiotics should be almost unknown. Molecules that are rather large or have several polar groups are solids at room temperature. Usually the antibiotics have several polar groups, which are involved in the interaction with bacterial macromolecules resulting in inhibition of bacterial growth. We must therefore accept this as the reason that even the smaller antibiotic molecules are solid substances.

The relationships between the chemical structures of the antibiotics and their activities are discussed in Chapter 5.

III. Producing Microorganisms

The large variety of molecules, mentioned in the preceding section, is produced by an array of widely diverse microorganisms, but the taxonomic distribution of the strains that produce them is neither uniform nor random. About 80% of the antibiotics described are in fact produced by strains that are members of only one bacterial order, the Actinomycetales, and especially by one genus of this order, the *Streptomyces*. The eubacteria very rarely produce antibiotics, except for those sporogenic bacilli that produce a particular class of polypeptide antibiotics. The fungi are frequently producers of antibiotics, but their structures exhibit less chemical variety than those found in the antibiotics from the actinomycetes.

The production of antibiotics is not rigorously species specific. The same antibiotic can be produced by organisms of species or genera or even orders that are different. And the reverse is also true; that is, strains classified taxonomically as members of the same species can produce different antibiotics. However, as a general rule, the more distant the organisms are on the taxonomic scale, the less probable it is that they will produce the same antibiotics.

The relationship between the producing microorganisms and the antibiotics produced is discussed in Chapter 9.

IV. Biosynthesis

In contrast with the great variety of chemical structures and of producing strains, the biological reactions involved in synthesis of the antibiotics can be grouped into a few fundamental biosynthetic pathways.

It is important to recognize that the biosynthetic pathways for the antibiotics are simple variations of the biosynthetic pathways of normal cellular metabolism, and it is surprising that small changes in these pathways can give rise to such diverse substances.

Because of the large size of the molecules involved, the substrate-enzyme specificity on occasion appears to be less rigid in antibiotic biosynthesis than in other biochemical reactions. Thus, a given enzyme may catalyze the same reaction in the presence of several slightly different substrates. On the other hand, the same intermediate can serve as the substrate of several enzymes. This partial lack of specificity results in the synthesis of products with a common basic structure but differing, for example, in degree of oxidation or unsaturation, or in other factors. For this reason, the antibiotics are often produced in families, which is to say that the same strain makes two or more antibiotics that resemble each other. On the basis of biosynthetic mechanisms, the antibiotics can be grouped as follows:

1. Analogs of primary metabolites (analogs of amino acids, nucleosides, coenzymes). These are small molecules that are biosynthesized in a manner similar to that for the primary metabolites and that often resemble them structurally.
2. Antibiotics derived by polymerization. These include: (a) polypeptide antibiotics and their derivatives, produced from the condensation of some amino acids to form a polypeptide chain that can then be modified by further reactions (it is important to note that the condensation of the amino acids does not take place by the classic protein synthesis mechanism); (b) antibiotics derived from acetate and propionate: There is a wide variety of chemical structures, but all are derived from reactions that follow the biosynthetic pathways for the fatty acids; (b) terpenoid antibiotics derived from isoprene synthesis (these are produced only by fungi and not by bacteria or actinomycetes); (d) aminoglycoside antibiotics, derived from condensation of sugar molecules, frequently amino sugars, and a cyclic amino alcohol (amino cyclitol).

In addition, some biosynthetic pathways have been described that are not similar to the general pathways, and antibiotics are known that derive from condensation of subunits originating from more than one of the pathways mentioned above.

The principle biosynthetic pathways are described in Chapter 6.

V. Activity and Resistance

The antibiotics are frequently grouped according to their *spectra of activity,* that is, according to the classes of microorganisms they inhibit. There are, therefore, *antiviral, antibacterial, antifungal, and antiprotozoal antibiotics.*

There are also *antitumor antibiotics,* products of microbial origin that inhibit the growth of cancer cells. The use of the term "antibiotic" is justified because these products were originally isolated on the basis of their antimicrobial activities.

The sensitivity of the different bacteria to antibiotics depends in large part on the structure of their cell walls, because this determines the capacity of the antibiotic to penetrate the bacterial cell. Therefore, the antibacterial antibiotics can be divided according to activity against Gram-positive or Gram-negative bacteria or mycobacteria. There are many more antibiotics effective against the Gram-positive bacteria, which are easily permeable. The antibiotics are said to have a *narrow spectrum of activity* if they are active only against Gram-positive bacteria, and a broad spectrum of activity if they are active against both Gram-positive and Gram-negative bacteria.

If the growth of a bacterial culture is inhibited by a given concentration of an antibiotic, it indicates that all the cells that constitute the bacterial population are sensitive to that antibiotic. However, in all bacterial populations, if they are large enough, some individuals are present having different characteristics from those of the originating strain because of chance mutation. These different characteristics may manifest themselves as biochemical differences (such as the capacity to synthesize an amino acid) or morphological differences, or, in what interests us, the sensitivity to an antibiotic. Mutants able to grow in the presence of a concentration of an antibiotic that inhibits the normal members of the population are said to be *resistant* to that antibiotic. The frequency of resistant mutants in different bacterial populations varies according to the microorganism and the antibiotic, and ranges between one mutant in every 10^7 cells to one in every 10^{10} cells. In some cases these mutants are resistant to minimal inhibitory concentrations of the antibiotic, but sensitive to slightly higher concentrations. In other cases they remain resistant to very high antibiotic concentrations. These two different situations have been given the names of *multistep* resistance and *single-step* resistance, respectively.

If the bacterial culture being treated with antibiotic contains some resistant cells, all the sensitive cells will stop growing, the resistant ones will continue to multiply, and within a short time the entire population will consist of the resistant mutants. This phenomenon is called *selection* of resistant mutants. However, it should be noted that in many cases the resistant mutants have some selective disadvantages as compared with the original sensitive type, so that in the long-term absence of antibiotic there is a tendency for the resistant population to revert to sensitivity through back mutation. If there are two antibiotics with similar structures and action and the bacteria that are resistant to one are found also to be resistant to the other, the two antibiotics are said to be *cross-resistant* to each other.

In recent years the picture of bacterial resistance to antibiotics has become much more complex, since some bacteria have been found to be able to transfer the property of resistance to given antibiotics to other bacteria of the same species and even of different species. This transfer is part of the more general phenomenon of exchange of genetic material from one cell to another that takes place by different mechanisms in different groups of microorganisms. The result is that in the presence of an "infectious" cell a microbial population can become resistant to an antibiotic without going through the normal selection process. This is known as *transferable* resistance.

The activities of antibiotics are discussed in more detail in Chapter 2 and the phenomenon of resistance in Chapter 4.

VI. Mechanism of Action

Antibiotics block the growth of sensitive microorganisms by inhibiting the action of a macromolecule, such as an enzyme or a nucleic acid, essential to the function of the cell. At the molecular level this means that the antibiotic molecule is able to bind to a specific site on the target macromolecule, forming a molecular complex that is no longer functional.

To determine the mechanisms of action of an antibiotic, one identifies the target macromolecule and its function. It is usually easier to identify the function that is blocked than the particular macromolecule involved, and for this reason we speak of antibiotics that inhibit synthesis of the cell wall, protein or RNA, or the replication of DNA, or membrane function, depending on what appears to be the primary effect of the antibiotic on the cell.

Some antibiotics are antimetabolites, acting as competitive inhibitors. These are structurally similar to normal metabolites, such as amino acids or coenzymes, and bind instead of these to the enzyme for which the metabolite is substrate or cofactor, thus inactivating it.

The selectivity of the actions of the antibiotics and thus the reasons why only some types of cells are inhibited are usually related to their mechanisms of action. Both of these aspects are examined in more detail in Chapter 3.

VII. Chemotherapy

Chemotherapy, the treatment of infectious diseases by administration of drugs, is based on the property of the antibiotics and of some other substances to inhibit the multiplication of the infecting microorganism through a selective toxic effect, without also having a toxic effect on the mammalian cells. This inhibition makes it easier for the body's defenses to overcome the infection. We have already stated that at least 3,000 antibiotics have been isolated, and tens of thousands of compounds have been synthesized chemically and tested for antimicrobial activity. Of all of these, only a few have been found to possess the characteristics necessary for clinical use, which can be summarized as follows:

1. *Activity against one or more pathogenic organisms.* Either a broad-spectrum or a narrow-spectrum antibiotic can be used once the organism has been isolated and identified and is known to be sensitive to one or the other. It is desirable to have one with low frequency of resistant mutants, and it is preferable, though not essential, that it also have bactericidal effect.
2. *Good absorption and distribution.* To be effective an antibiotic must be

absorbed, reach the site of the infection, and remain there in concentrations greater than the minimal inhibitory concentration (MIC) for a sufficient length of time. It must be eliminated from the body to avoid accumulation.

3. *Lack of toxicity*. The antibiotic must be without any significant toxicity to the host at the doses used in therapy. Some adverse reactions, if they are slight and not very frequent, are not obstacles mitigating against the use of the drug. More serious adverse reactions are tolerable only if the antibiotic is to be used in diseases that are extremely serious or potentially lethal.

The concepts and the principal methods used for determining these properties for new antibiotics are described in Chapter 7. Clinical use of the antibiotics is briefly discussed in Chapter 8.

VIII. Chemical Modifications

In relatively complex molecules such as those of some antibiotics, some structural components or chemical groups are directly involved in the formation of the complex with the macromolecule that is its target. Other structural components or groups in the molecule are not directly involved in this and can be modified chemically without substantially modifying the intrinsic activity.

Moreover, by these alterations one can modify some physicochemical characteristics of the molecule, especially those involved in water or lipid solubility, and this is extremely important with respect to the pharmacokinetics, and hence the possibility for use in therapy. These same properties, such as solubility in water or lipid, may also affect the spectrum of activity, as they influence the penetration of the cell.

These are the reasons for the large-scale effort to chemically modify the natural antibiotics that have been of great importance in recent years in providing new products for therapeutic use.

IX. Principal Classes of Antibiotics

Various schemes for classification of the antibiotics have been proposed, none of which has been universally adopted. At present, those natural or semisynthetic antibiotics that have a common basic chemical structure are grouped into one "class," and called by a name derived from the first one to be known or by one that refers to a principal chemical property. This type of empirical classification is very useful in practice, because the components of one class usually have many biological properties in common, as is evident from the brief descriptions in the following sections. A more detailed description of the properties of the different classes of antibiotics is given in Chapter 5.

A. β-Lactam Antibiotics (Penicillins and Cephalosporins)

The penicillins were the first antibiotics to be used in therapy and are still considered the drugs of first choice for treatment of many infections. They and the more recently developed cephalosporins comprise the β-lactam group, so called because of the presence in their molecules of a four-atom ring with the chemical name β-lactam. They are synthesized by fungi of the genera *Penicillium* and *Cephalosporium*. Recently actinomycetes of the genera *Streptomyces* and *Nocardia* and also organisms of the genus *Pseudomonas* have also been shown to produce β-lactam antibiotics. From the biosynthetic aspect, they may be considered to be derived from polymerization of amino acids.

The β-lactam antibiotics inhibit the synthesis of peptidoglycan (see Chapter 3, section II), a basic component of the bacterial cell wall, causing irreversible damage. They are therefore bactericidal antibiotics. They are not active against fungi, whose cell walls do not contain peptidoglycan, or against mycoplasma, which lack cell walls.

The spectrum of action of the initial penicillins, such as the widely-known penicillin G, was limited to Gram-positive bacteria and Gram-negative cocci.

By means of chemical modifications, derivatives have been obtained that are effective in varying degrees against almost all of the Gram-negative bacteria. The same results have been obtained with the cephalosporin derivatives starting from the weakly active original cephalosporin C. Other objectives obtained by semisynthesis are the preparation of penicillins and cephalosporins that are active when given orally and of derivatives somewhat less sensitive to the enzymes, the β-lactamases, which are drug-inactivating enzymes produced by some resistant bacterial strains. Several new β-lactams produced by streptomycetes and their semisynthetic derivatives are now actively investigated. Since they do not belong to the penicillin or cephalosporin groups, they are referred to as "nonclassic" β-lactams.

With certain exceptions, the toxicity of the penicillins and cephalosporins is very low. The major problem in the use of these antibiotics is the appearance of hypersensitization phenomena and allergy, sometimes with very severe manifestations.

B. Aminoglycoside Antibiotics

The aminoglycoside antibiotics are a large class of substances produced by members of the genera *Streptomyces, Micromonospora,* and *Bacillus.* They are characterized chemically by the presence of a cyclic amino alcohol to which some amino sugars are bound. Both the amino alcohol and the amino sugars are derived biosynthetically from glucose. The aminoglycoside antibiotics irreversibly inhibit protein synthesis by interacting with ribosomes, which is a bactericidal effect. They are particularly active against Gram-negative bacteria. Streptomycin, the first one known in the class, was in fact the result of a research program aimed at isolating an antibiotic effective against Gram-negative organisms.

It was also the first antibiotic that was found to be effective against tuberculosis. Later intensive research gave rise to, among others, gentamycin, tobramycin, kanamycin, and amikacin, which are active against bacterial classes not very sensitive to streptomycin.

Because of their chemical structure, all these antibiotics are very water-soluble and are, therefore, not absorbed when given orally.

Their major adverse effects are damage to ear nerves and to the kidney.

C. Tetracyclines

The tetracycline family is important because of its very broad spectrum of action and its great therapeutic effectiveness. It originally included chlorotetracycline, oxytetracycline, and tetracycline, the last being the most widely used clinically.

They are products of different strains of *Streptomyces.* They are biosynthesized by cyclization of a chain obtained by condensation of acetate and malonate units. The chemical structure consists of four rings condensed linearly, and this is the basis of their name.

They act by preventing ribosomal protein synthesis. The effect is reversible and, therefore, they are bacteriostatic agents. Their activity spectrum is particularly broad, including Gram-positive and Gram-negative bacteria, rickettsiae, chlamydiae, and some protozoa. Because of their physicochemical properties (they are insoluble at neutral pH), the natural tetracyclines can only be given orally. A great deal of study of semisynthetic products resulted in the preparation of injectable derivatives and derivatives with prolonged action.

Only limited success has been achieved in the search for derivatives with activity against resistant bacteria.

The tetracycline antibiotics are well tolerated at the doses needed for effectiveness. The greatest limitation to their usefulness consists at present of the wide occurrence of resistant strains.

D. Macrolides

The macrolides have chemical structures containing a ring consisting of no fewer than 12 carbon atoms and closed by a lactone group. They are typical products of the streptomyces. They appear to be biosynthesized by condensation of acetate or proprionate units.

The macrolides can be subdivided into two rather homogeneous classes:

1. *Antibacterial macrolides,* characterized by the presence of lactone rings of 12–16 atoms, with at least two sugar molecules attached. They reversibly inhibit protein synthesis by interacting with the ribosomes. They are bacteriostatic and have an activity spectrum that for all practical purposes is restricted to Gram-positive bacteria.

A typical representative of this class is erythromycin. Derivatives of this antibiotic have been synthesized to improve its oral absorption, which is quite irregular. Some derivatives for intramuscular administration have been prepared, but with unsatisfactory results.

2. *Antifungal and antiprotozoal macrolides,* characterized by lactone rings with 30 or more atoms, with hydroxyl substituents and including a series of 4–7 conjugated double bonds. For this last reason, they are called polyenes (tetraenes, pentaenes, etc.). They are active only when given intravenously. They induce distortions in the cell membranes by interfering with the sterols. Therefore, they are not active against bacteria, which do not contain sterols in the cytoplasmic membrane, but only against fungi and some protozoa. Their toxicity makes it necessary to limit their use to the most serious cases. The best known representative of this family is a heptaene, amphotericin B.

E. Ansamycins

These are the newest of the antibiotics used for human therapy. The typical structural characteristic of the ansamycins is an aliphatic chain that connects two opposite points of an aromatic ring, like a handle or ansa (hence the name). They are produced by strains of several genera of the order Actinomycetales. Their biosynthesis resembles that of the macrolide antibiotics.

The ansamycins can be divided into two subgroups, the naphthalenes and the benzenes, according to the type of aromatic ring present. The naphthalene ansamycins are antibacterial, selectively inhibiting the enzyme RNA polymerase. The benzene ansamycins are less selective in their action, and some have been studied as possible antitumor agents. (One of these is maytansin, which curiously enough is a plant product.)

Among the naphthalene ansamycins are the rifamycins, which are very active against Gram-positive bacteria and mycobacteria. The natural rifamycins are not used in therapy but, instead, some semisynthetic rifamycins have chemotherapeutical roles, such as rifamycin SV, which is used against infections of the bile tract, and rifampin, which is active orally and has a broad spectrum of action, being particularly effective in the treatment of tuberculosis.

F. Polypeptide and Depsipeptide Antibiotics

The polypeptide antibiotics are comprised of chains of amino acids, often closed into a ring. They are produced by a great variety of microorganisms by biosynthetic pathways different from those in which amino acids are polymerized to give proteins.

Some of these, such as bacitracin and gramicidin, are of only historical interest, since they are too toxic to be given systemically.

Among the most important of these in use are the polymyxins, which are very effective against Gram-negative bacteria and, therefore, useful in severe urinary tract infections even though they are rather toxic. Many polypeptide antibiotics act by interfering with the structure of the cytoplasmic membrane and, hence, are prevalently bactericidal. Other polypeptide antibiotics interfere with the process of protein synthesis.

One polypeptide antibiotic with a particularly complex structure is bleomycin, which is used as antitumor agent because of its capacity to cause rupture of DNA chains.

The depsipeptide antibiotics are comprised of chains of amino acids alternating with oxyacids. One typical representative of this class is valinomycin, an "ionophoric" antibiotic, which acts by interfering with the transport of potassium ions into the cells.

G. Miscellaneous Antibiotics

Some antibiotics are used in therapy that cannot be classified in any of the families so far described. Some of these are:

1. *Chloramphenicol,* originally isolated from a streptomycete, is now produced synthetically since it has a relatively simple structure. It is one of the very few natural compounds that contains a nitro group. It inhibits protein synthesis and is bacteriostatic. It is very effective in the cure of infections with Gram-negative organisms, especially in typhoid fever. It is active orally.
2. *Lincomycin* resembles to a large extent erythromycin in its mechanism of action (inhibition of protein synthesis) and its spectrum of action (limited to Gram-positive organisms), and in fact it demonstrates a partial cross-resistance to erythromycin.

   Chemically, it is completely different, being made up of a modified amino acid condensed with a complex amino sugar. It is particularly active against anaerobic bacteria. Clindamycin, a semisynthetic derivative, has similar properties.

3. *Fusidic acid* is produced by a fungus. It has a steroid structure and is active against Gram-positive organisms. It is an inhibitor of protein synthesis, and it is active orally and only very slightly toxic. The principal limitation in its usefulness is the high frequency of resistant mutants, for which reason it is usually given in combination with other antibiotics.

4. *Vancomycin* is a complex product of *Streptomyces orientalis,* having bactericidal activity against Gram-positive bacteria. It acts by interference with the synthesis of the cell wall. It does not give rise to development of resistance. It is used only for serious illnesses, because it is effective only when given intravenously.

5. *Griseofulvin* is one of the very few antifungal drugs that can be given systemically. It is made by a fungus and has an aromatic structure derived biosynthetically by condensation of acetate and malonate units. It is particularly effective in dermatophyte infections.

6. *Daunomycin and adriamycin* are two very similar antibiotics produced by a streptomycete chemically belonging to the family of the anthracyclines. They are cytostatic and form complexes with DNA. Therefore, they are used in the treatment of various types of tumors.

Chapter 2

Activity of the Antibiotics

I. Definition

The activity of an antibiotic is defined and measured in terms of its capacity to inhibit microbial growth (bacteria, fungi, protozoa). Although the concept of growth is familiar as applied to macroscopic organisms (irreversible increase in volume as a function of time), it must be redefined to take into account microscopic organisms, because in this case two different levels of growth can be distinguished: population growth and growth of a single cell. By population growth we mean the increase with time of the number of microorganisms, i.e., the increase in the density of the microbial population. By cell growth we mean the synthesis of cellular material needed for one cell to give rise to two daughter cells. Obviously, the growth of the microbial population is the result of cell growth.

The activity of an antibiotic is determined as its capacity to inhibit the growth of the microbial population and is measured either by directly counting the number of cells per unit volume or by evaluating some parameter of the culture that is correlated with population density, such as the property of light diffraction.

In the following sections we refer to antibacterial activities, but the same techniques can be adapted to measurement of antifungal or antiprotozoal activity.

II. Determination of the Minimal Inhibitory Concentration in Liquid Medium

One method to quantify the activity of an antibiotic is to determine the minimal concentration needed to inhibit completely the growth of a certain bacterium. This is called the minimal inhibitory concentration (MIC).

The MIC is determined as follows:

1. A series of test tubes is prepared, all containing the same volumes of medium inoculated with the test bacterium (from 10^3 to 10^6 bacteria per milliliter).

2. Decreasing concentrations of antibiotic are added to the tubes. Usually stepwise dilution by a factor of 2 is used (i.e., if the concentration of antibiotic in the first tube is 1 mg/ml^{-1}, in the second tube it will be 0.5 mg/ml^{-1}, and in the third 0.25 mg/ml^{-1}, and so on). One tube is left without antibiotic, to serve as a positive control for bacterial growth.

3. The cultures are incubated in an oven at a temperature that is optimal for the test bacterium and for a period of time sufficient for the growth of at least 10–15 generations (usually overnight).

4. The tubes are inspected visually to determine where bacteria have grown, as indicated by turbidity. Turbidity of the culture medium is interpreted as the presence of a large number of bacteria ($>10^7$/ml^{-1}). The tubes in which the antibiotic is present in concentrations great enough to inhibit bacterial growth remain clear (Fig. 2.1). In experimental terms the MIC is the concentration of antibiotic present in the last "clear" tube.

Obviously, because of the way it is determined (stepwise 2-fold dilutions of the antibiotic and visual determination of growth), the MIC is not a very precise value. In fact, displacement by only one tube in determination of growth causes a 100% variation in the MIC. However, the MIC is a very useful parameter both for study of the biology of the antibiotic and for its clinical use. *The MIC is a characteristic of a particular antibiotic for a particular bacterial species under particular test conditions.* In addition, within a given species, different strains may be, within certain limits, more or less sensitive to the antibiotic and therefore may have different MIC values.

III. Determination of the Minimal Inhibitory Concentration in Solid Medium

In concept this is similar to the determination of the MIC in liquid medium. One prepares a series of Petri dishes with stepwise dilutions of antibiotic dissolved in the culture medium. Melted agar is added and the plates are allowed to cool and harden. The surfaces are then streaked with cul-

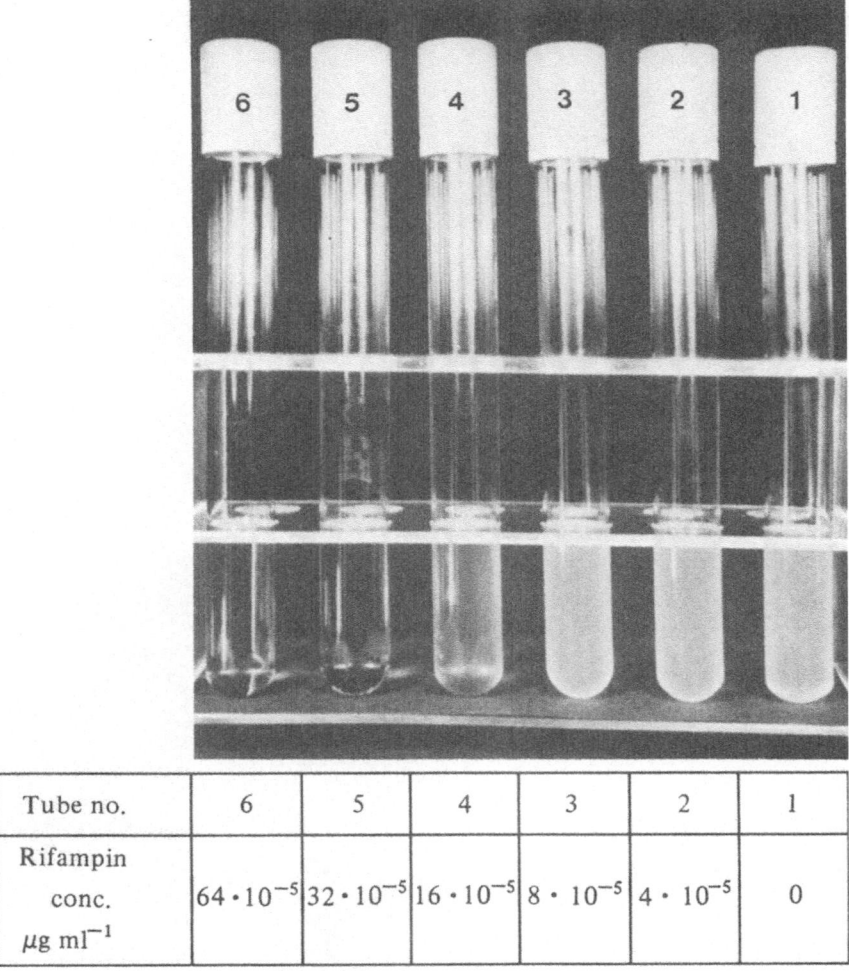

Tube no.	6	5	4	3	2	1
Rifampin conc. $\mu g\ ml^{-1}$	$64 \cdot 10^{-5}$	$32 \cdot 10^{-5}$	$16 \cdot 10^{-5}$	$8 \cdot 10^{-5}$	$4 \cdot 10^{-5}$	0

Figure 2.1. Determination of the MIC of rifampin against *Staphylococcus aureus* in liquid medium. Each tube contains 5 ml of microbiological medium inoculated with 10^3 bacteria per milliliter. The values below each tube indicate the concentrations of rifampin (tube 1 contains no antibiotic and serves as the positive control for growth). The MIC is the concentration present in tube 5 (0.00032 $\mu g/ml^{-1}$).

tures of one or more bacteria, and after a suitable incubation period one can determine the minimal concentration that inhibited formation of the bacterial film. One advantage over the liquid medium method is that a single plate can be seeded in different areas with different species or strains of bacteria, enabling the MIC for several bacteria to be determined in a single operation (Fig. 2.2).

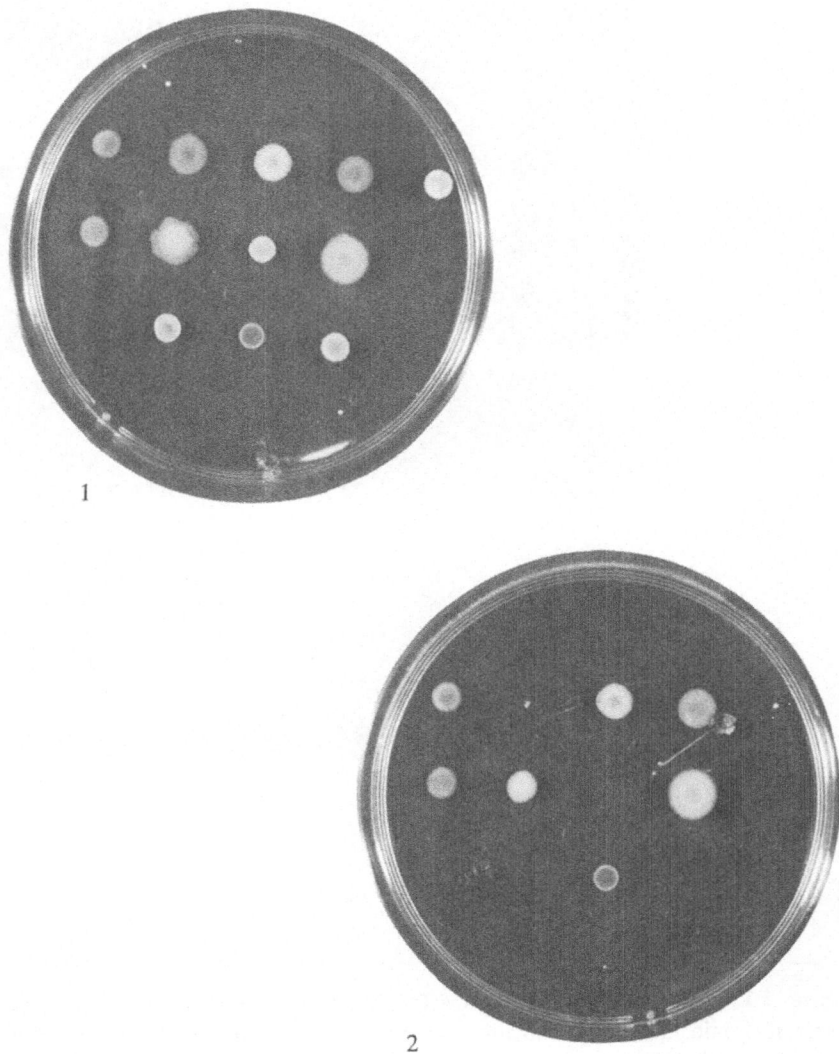

Figure 2.2. Simultaneous determination of the MIC for rifampin against 12 species of bacteria in solid medium. Plate 1 is the positive control and contains no antibiotic. In plates 2, 3, and 4 there are increasing concentrations of antibiotic, 0.02, 0.04 and 0.08 μg/ml^{-1}. One can easily see that different bacteria have different sensitivities to the antibiotic. With a sufficiently large number of dilutions, the MICs for all 12 of the bacterial species can be determined at the same time.

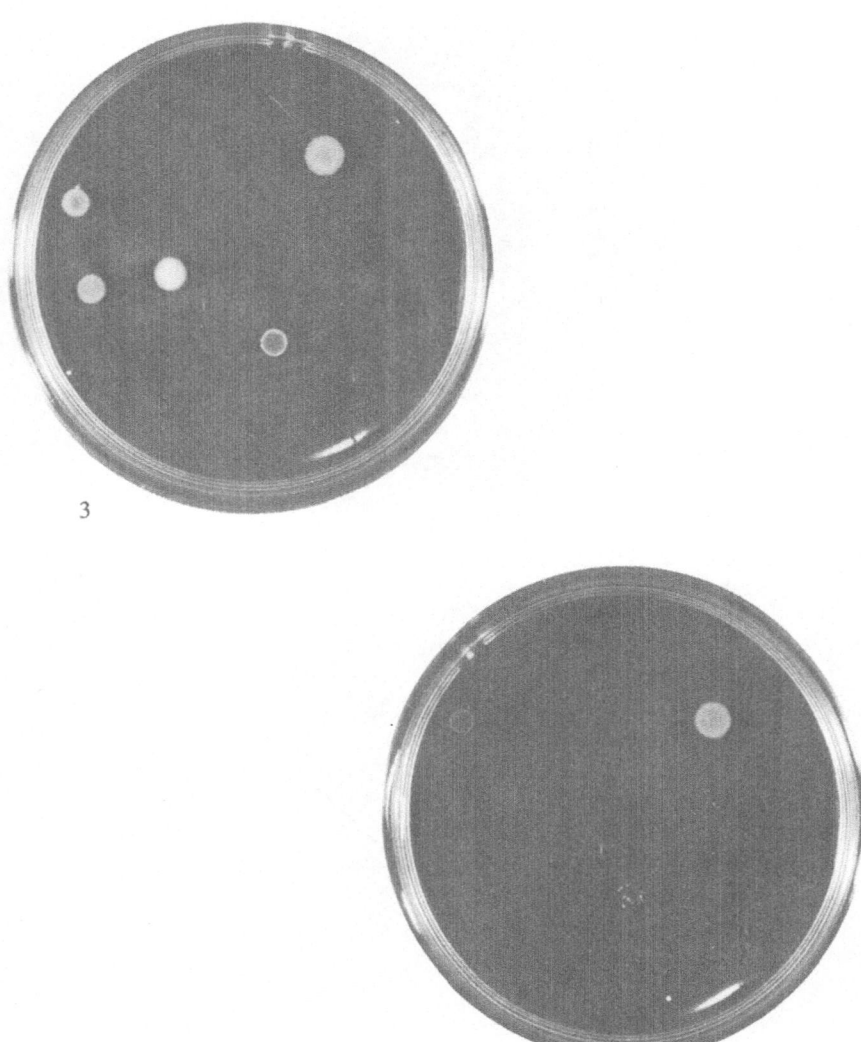

3

4

IV. Determination of Antibiotic Activity by the Agar Diffusion Test

This is a suitable method for determination of the concentration of an antibiotic in a solution. It consists of placing filter paper discs soaked with the antibiotic solution, which is to be assayed on the surface of an agar medium containing a dilute suspension of bacteria. After an appropriate incubation period, the surface of the agar, which had been transparent, acquires a turbid appearance as a result of the light diffraction caused

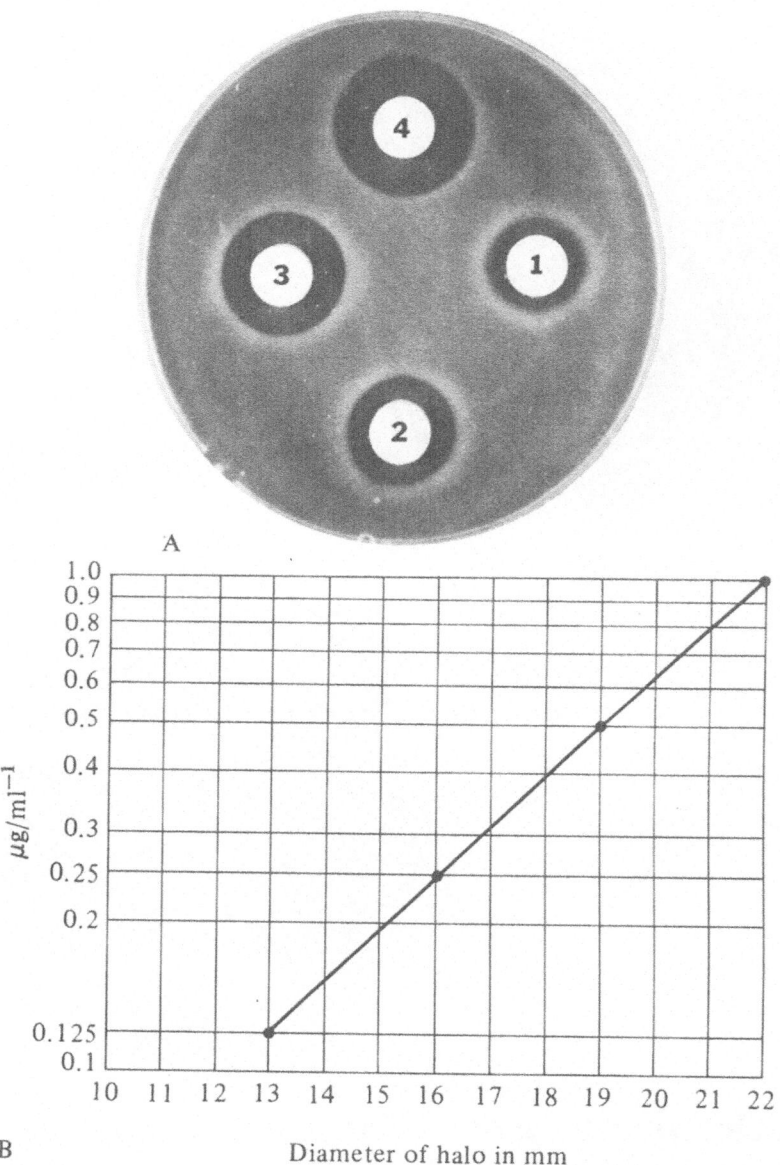

Figure 2.3. Construction of a standard curve for the inhibition of *Staphylococcus aureus* by rifampin: **A.** Discs 1, 2, 3, and 4 were soaked with equal volumes of rifampin solutions with concentrations of 0.125, 0.25, 0.5, and 1 μg/ml^{-1} and were then placed on the surface of agar seeded with *Staphylococcus aureus*. The relationship between the logarithms of the concentrations of the solutions in which the discs had been immersed to the diameter of the inhibition halo is shown in the graph in **B**. Diameter of halo is given in millimeters.

by the bacterial growth. However, transparent haloes will remain around the filter paper discs because the antibiotic has diffused into the agar, therefore inhibiting bacterial growth. When carried out under standardized and rigorous test conditions, the diameter of the halo of inhibition is a function of the logarithm of the concentration of the antibiotic. A standard curve can be prepared with known antibiotic concentrations, and the concentrations in the unknown solutions can be read from it (Fig. 2.3).

On occasion small cylinders with capillary openings or small wells dug in the agar and filled with antibiotic solution are used instead of the filter paper discs.

V. Factors Affecting the Determination of Antibiotic Activity

The in vitro activity of an antibiotic is affected by the test conditions under which it is determined (Table 2.1). Some of the relevant factors are: the composition of the medium, the density of the bacterial inoculum, the total number of bacterial cells in the inoculum (size of the inoculum).

A. Composition of the Medium

Let us take as a simple example an antibiotic that acts by inhibiting the biosynthesis of an amino acid. If it is tested in a medium without that amino acid it will appear to be very active; i.e., it will have a low MIC. If it is tested in a medium containing the amino acid and the bacterium can take up the amino acid from the medium, the antibiotic will appear to be inactive.

In addition to such specific effects, one often encounters less specific effects of the culture medium, that is, those not directly correlated with the mechanisms of action of the antibiotic or with its chemical structure. One particularly interesting example is the presence of serum. It is introduced into the medium in determination of the MIC to reproduce physiological conditions similar to blood. Many antibiotics bind to serum proteins

Table 2.1. Factors Affecting the Determination of Activity of Antibiotics

Testing organism
Medium composition (pH, ions, serum, antagonists)
Inoculum size and inoculum density (etherogenicity of population, detoxifying mechanisms, carryover of antagonistic substances)
Incubation conditions (time, temperature, aeration)

(essentially to the albumin) and this decreases the number of free molecules that are available to enter into the bacterial cell. The binding of an antibiotic to serum proteins is usually correlated with the lipophilicity of certain of its chemical substituents.

Obviously, the conditions for determining the MIC in liquid medium differ from those in the case of solids, if for no other reason than that solid medium contains agar. Agar with its SO_3-groups can absorb antibiotics, thus changing the conditions for diffusibility of the antibiotic, of dissolved oxygen, and of nutrients. It is not surprising therefore that the MIC for an antibiotic against a given bacterium will differ according to whether it is determined in liquid or solid medium, even when the two media are the same except for the agar.

It must also be kept in mind that the physiology of the cells may differ when they grow individually in liquid and when they grow in colonies on a solid surface.

A very important effect on antibiotic activity is that of the pH of the culture medium. In addition to minor effects, as, for example, the pH effect on the microorganism's growth rate, and hence indirectly on its susceptibility, the pH also has a very large and direct effect on the ability of the substance to penetrate the bacterial cell. For example, nonionized substances diffuse better through the cell wall and cytoplasmic membrane than do ionized substances. Therefore, the pH of the medium, by affecting the degree of ionization of a basic or acidic antibiotic, can influence directly its rate of penetration into the bacteria and hence its effectiveness.

B. Density and Size of Bacterial Inoculum

The density of the inoculum is the number of bacteria inoculated divided by the volume in which they are tested, usually expressed as number of cells per milliliter of culture. The size of the inoculum is the total number of bacteria inoculated. The MIC of many antibiotics is not affected by variation in the inoculum density in the range commonly used (10^3–10^6 bacteria/ml^{-1}).

In fact, even when one uses very low concentrations of antibiotic, for example, 0.01 $\mu g/ml^{-1}$, the ratio of the number of molecules to the number of bacteria is very high (0.01 $\mu g/ml^{-1}$ of an antibiotic with a molecular weight of 1000 contains $6 \cdot 10^{12}$ molecules per milliliter). However, there are some exceptions. For example, quite often a large number of the antibiotic molecules are adsorbed on the outer surface of the bacterial cell, reducing the number of free molecules available to enter the bacterium to an insufficient number if the bacterial density is high. In addition, often a large number of antibiotic molecules is necessary to inhibit the growth of a single cell. At times the bacteria produce and excrete into the culture medium enzymes that can destroy the antibiotic (such as β-lactamases, which can destroy β-lactam antibiotics; see Chapter 4). The amount of antibiotic that is destroyed is essentially a function of the concentration of the enzyme in the culture medium, and this depends on the inoculum.

At first glance it would appear not to make any difference, once the density of the bacteria has been fixed, whether the test is carried out in small volumes, as, for example, 0.25 ml or less in the miniaturized systems, or in the traditional volume of 10 ml in common laboratory test tubes. It would not make any difference if all the members of the bacterial population were identical, if there were not always a certain variability in susceptibility from cell to cell.

When the total number of bacteria in the inoculum is very high, there is greater probability that some of the cells will be less susceptible to the antibiotic. All the susceptible cells will be inhibited, but the less susceptible ones (in theory, even a single cell) will multiply and, after the usual 18 h of incubation, there will be a dense population of bacteria. A large variation in MIC as a function of the number of cells inoculated usually indicates a high frequency of mutants resistant to the antibiotic (see Chapter 4). The frequency with which resistant mutants appear is different for different antibiotics. For those of clinical interest, it varies from 10^{-7} to 10^{-10}.

The factors we have described in this section are only some of the many that can affect the activity of an antibiotic against a given bacterium (see Table 2.2). Therefore, it is indispensable for all test conditions to be specified precisely and, if possible, standardized, if one wishes to use inhibition of bacterial growth for *quantitative* determination of the activity of an antibiotic and to have data that are reproducible in different laboratories.

VI. Determination of the Minimal Bactericidal Concentration

Inhibition of growth of a bacterial population by an antibiotic can be the result of the inhibition of the capacity of each individual cell to duplicate itself or it can be the result of all the cells having been killed. In the first situation, if one removes the antibiotic, the bacterial culture should start to grow again, which will, of course, not occur in the second situation. The antibiotic's action in the first case is called bacteriostatic, and in the second, bactericidal. A single antibiotic may be both bacteriostatic and bactericidal, depending on the concentration used. Sometimes it is useful to determine the minimal bactericidal concentration (MBC). This is done as follows: (a) proceed as in the determination of MIC in liquid medium; (b) at the end of the incubation, take an aliquot of liquid from each tube in which no growth can be seen, dilute it to remove the antibiotic, and inoculate the surfaces of Petri plates. These are incubated for 24 h at an

Table 2.2. Factors Affecting Activity of Some Representative Antibiotics

Antibiotic	Factors
Aminoglycosides	pH; divalent cations (Mg^{2+}, Ca^{2+})
Penicillins	Inoculum size
Tetracycline	pH; divalent cations (Ca^{2+})
Polyene macrolides	Sterols
Trimethoprim	Thymine, glycine, methionine
Polyether antibiotics	Monovalent ions (K^+)
Methicillin	Inoculum size (uneven distribution of susceptibility)
Sulfonamides	p-Aminobenzoate, inoculum size (carryover of antagonists), one-carbon metabolism products

appropriate temperature and colony growth is observed. The MBC is the concentration of antibiotic that was present in the first culture tube in the series from which no colonies could be grown on the Petri plates. In practice, "no colonies" is defined as a 99.9% reduction in colony forming units (CFU).

VII. The Antibiogram (Bacterial Sensitivity Test)

To carry out a rational treatment with antibiotics, it is always useful and often essential to know to what antibiotics the pathogenic bacterium causing the infection is susceptible and to what antibiotics it is resistant.

The name antibiogram has been given to the complex of information about the susceptibility or resistance of a microorganism to a series of antibiotics.

Usually the antibiogram is determined as follows: The microorganism is isolated from the patient and grown in pure culture, after which agar plates are inoculated with it and small discs containing standard concentrations of the commonly used antibiotics are placed on them. The sizes of the inhibition haloes indicate the degrees of susceptibility (or resistance) of the microorganism to the different antibiotics (Fig. 2.4).

VIII. Spectrum of Antibiotic Activity

Occasionally, an antibiotic is active against all microorganisms, but usually different antibiotics are effective against different organisms. The group of microorganisms whose growth is inhibited by an antibiotic make up its spectrum of activity.

This is expressed quantitatively by the MIC values for a series of micro-

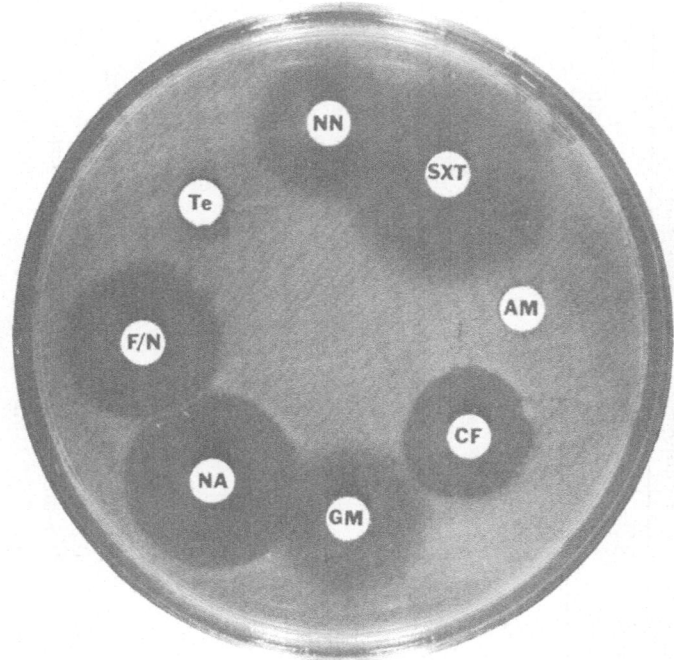

Figure 2.4. Antibiogram. The microorganism is *Escherichia coli* which can be seen to be sensitive to cephalothin (*CF*), gentamycin (*GM*), nalidixic acid (*NA*), nitrofurantadin (*F/N*), tobramycin (*NN*) and co-trimoxazole (*SXT*). It is resistant to tetracycline (*Te*) and ampicillin (*AM*).

organisms considered to be representatives of the various taxonomic groups that are most important from the clinical point of view. In Table 2.3 are shown the spectra of activity of some important antibiotics.

Antibiotics can be classified as antibacterial, antifungal, or antiprotozoan according to their spectra of activity.

The antibacterial antibiotics are those that inhibit the growth of prokaryotic microorganisms (bacteria). The antifungal antibiotics inhibit the growth of yeasts, molds, or fungi. The antiprotozoans inhibit the growth of such protozoa as trichomonas, amoeba, coccidia, or others. Some antibiotics are active against two, or even against all three, of these groups. However, these usually cannot be used clinically because they are too toxic.

The antibacterial antibiotics can be divided into two categories: (1) those that inhibit the growth of Gram-positive bacteria and are inactive against Gram-negative, and (2) those that inhibit both Gram-positive and Gram-negative organisms. These latter are often called broad-spectrum antibiotics. It is very seldom that an antibiotic is found that is active only against Gram-negative bacteria.

Table 2.3. MICs ($\mu g/ml^{-1}$) of Some Representative Antibiotics against Representative Pathogenic Microorganisms

Microorganism	Penicillin G	Ampicillin	Cephaloridine	Erythromycin	Tetracycline	Chloramphenicol	Streptomycin	Rifampin	Amphotericin B
Gram-positive Cocci									
Diplococcus pneumoniae	0.010	0.04	0.003	0.01–2	0.05–1	1–3	0.2–1	0.02	uns[a]
Staphylococcus aureus									
Penicillinase nonproducer	0.04	0.1	0.05	0.01–2	0.1	4–8	1–5	0.002	uns
Penicillinase producer	50	uns	13	0.01–2	0.1	4–8	1–5	0.002	uns
Streptococcus fecalis	1–4	1–5	10	0.2–3	0.1	2–5	2–5	0.01–10	uns
Streptococcus hemolyticus	0.008	0.02	0.001	0.01–0.3	0.05–5	0.7–2	2–5	0.01	uns
Gram-negative Cocci									
Neisseria gonorrheae	0.05	0.1–0.6	0.1–4	0.02–1	0.1–3	1	2–uns	0.02	uns
Gram-positive Bacilli									
Clostridium tetani	0.01–3	0.02	0.5	0.5–2	0.03–1	0.1–uns	uns	0.1	uns
Corynebacterium diphtheriae	0.3–3	—	1	0.1–3	0.5–5	0.5–5	0.5–uns	1	uns
Gram-negative Bacilli									
Bacteroides fragilis	—	50	uns	—	—	—	uns	0.005–0.2	uns
Escherichia coli	uns	2–uns	1–uns	2–10	0.5–5	2–12	1–uns	1–100	uns
Klebsiella spp.	6	1.5	6–uns	uns	0.5–uns	0.3–1.5	8–uns	5–10	uns
Hemophilus influenzae	0.5–3	0.25	8–uns	uns	0.2–5	0.2–5	1–5	0.02–2	uns

Table 2.3. (*continued*)

Microorganism	Penicillin G	Ampicillin	Cephaloridine	Erythromycin	Tetracycline	Chloramphenicol	Streptomycin	Rifampin	Amphotericin B
Gram-negative Bacilli									
Proteus mirabilis	16–32	uns	4–uns	uns	uns	4–16	4–uns	1–10	uns
Pseudomonas aeruginosa	uns	uns	uns	uns	4–200	16–uns	20–500	10–100	uns
Salmonella typhi	6	0.2–2	1–6	uns	0.5–10	0.2–5	8–16	20	uns
Shigella spp.	16	1–8	1–8	uns	0.5–uns	0.5–8	2–8	20	uns
Vibrio cholerae	uns	—	—	uns	0.5–10	1–uns	10–100	0.1–1	uns
Mycobacterium tuberculosis	uns	—	10	uns	uns	6–30	0.5–5	0.5	uns
Spirochetes									
Treponema pallidum	0.02–0.2	—	—	0.03	—	5	—	uns	uns
Miscellaneous									
Mycoplasma pneumoniae	uns	uns	uns	0.01	0.1–5	5	4–16	5–100	
Rickettsiae	uns	uns	uns	uns	0.1–1	1	uns	—	uns
Yeasts and Fungi									
Candida albicans	uns	uns	uns	uns	uns	uns	uns	uns	0.1–3
Dermatophytes spp.	uns	uns	uns	uns	uns	uns	uns	uns	0.2–40

^a uns, Unsusceptible.

A. Clinical Significance

The clinical significance of the spectrum of activity is obvious. One cannot hope to cure an infection due to Gram-negative bacteria by giving an antibiotic that is effective only against fungi or against Gram-positive bacteria. The MIC values of the spectrum of activity also provide preliminary information about the minimal concentrations of the antibiotic in the body fluids required for activity against the different pathogenic microorganisms.

B. Biological Significance

The existence of different spectra of activity for different antibiotics indicates that they act specifically, unlike many sterilizing or disinfecting agents that inhibit cell growth without any specificity. The phenomenon of the specificity of action of the antibiotics and the cytological and molecular bases for this are discussed in Chapter 3.

IX. Interaction of Antibiotics

The term *interaction* or *interference* refers to the effect that an antimicrobial agent has on the degree of inhibitory action of a second antimicrobial agent (and vice versa) when both are supplied simultaneously to a microbial culture (combination).

Synergism is assumed when the antimicrobial effect of the combination is greater than the sum of the effects of each component alone; *additivity* (or addition) is assumed when the effect of the combination is equal to the sum of effects of the single components. *Antagonism* is assumed when one component decreases the antimicrobial activity of the other. *Indifference* refers to a situation in which the presence of one component does not alter the activity of the other.

Although these four situations are conceptually clear and distinct, the quantitative determination of antimicrobial interaction is complex and the results are sometimes dependent on the method used and the nature of the parameter determined (end point inhibition: MIC or MBC, or variations in growth rate or killing rate at subinhibitory concentrations).

These interactions are best expressed graphically by the use of the isobologram, in which the concentration of drug A present in each inhibitory end point is plotted against the concentration of drug B that gives the same end point value. The shape of the curve indicates what type of interaction is taking place (Fig. 2.5). A less precise but simpler method to detect interaction consists in placing two paper strips each soaked in an

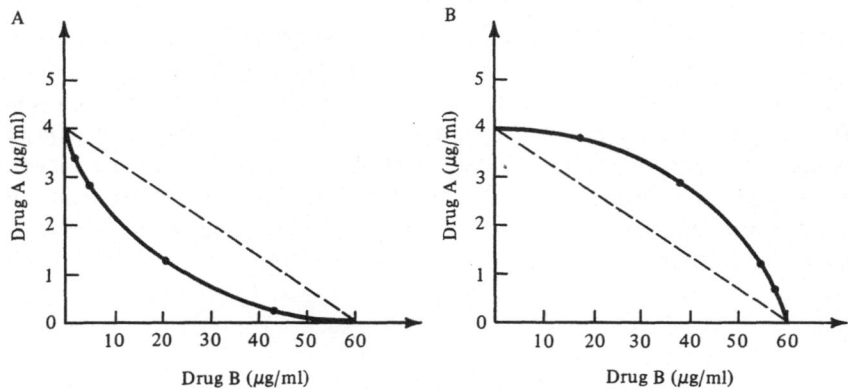

Figure 2.5. Isobolograms indicating synergism (**A**) or antagonism (**B**) between drugs A and B. The points on the right of the solid line represent the concentrations of drug A and B which together inhibit the growth of the microbial strain. Those on the left represent noninhibitory concentrations. The degree of synergism or antagonism is indicated by the shape of the curve with respect to the dotted line, which represents the theoretical situation of simple additivity.

antibiotic on the surface of an agar plate. The agents diffusing from the strip into the agar generate a concentration gradient. A dip in the angle of growth in the area of the intersection of the two strips is indicative of interaction.

X. Recent Microbiological Techniques: Miniaturization and Automation

The use of microbiological analytical techniques has grown considerably in recent years. Controls of sterility have become ever more frequent, not only in the pharmaceutical and food industries, but also in chemical and electronic component industries. In addition, the preoccupying phenomenon of the spread of bacteria resistant to more than one antibiotic has caused an enormous increase in the number of requests for antibiograms sent to the microbiological laboratories of the hospitals. As a consequence, attempts have been made to develop microbiological techniques that would be less labor intensive and less costly. These attempts have developed in two directions.

1. *Miniaturization of the tests.* Traditionally, tests have been carried out in 10 ml of medium, but at present tests are available and widely used that are carried out in 25–50 μl, which saves a great deal of material and labor.

2. *Automation.* Some automated systems are based on automatic programs for all the operations involved in the test (sample preparation, bacteria preparation, incubation, reading of the results and recording them). Others automate only the operations of dilution and incubation, while the bacterial growth is determined by traditional methods. Still others require manual operation of the initial steps and allow automation of the phases of reading and transcription of the results. These last systems provide the data in forms that can be directly fed into a computer, thus decreasing cost and time and the frequent errors of manual transcription.

Chapter 3

Mechanism of Action
of the Antibiotics

I. Relationship Between the Mechanism
of Action and the Selectivity
of the Antibiotics

An antibiotic inhibits growth of a microbial population. Population growth results from reproduction of individual cells, that is, from duplication of cellular material and subsequent division of the cell into two daughter cells. For an antibiotic to affect a microbial cell, it must (1) enter the cell, (2) bind physically to a cellular structure involved in some process essential for maintenance of the life or the growth of the cell, and (3) completely inhibit the process in which that structure is involved.

At the cellular level, the antibiotic effect can be one of the following: (1) bacteriostatic, that is, it blocks the ability of the cells to replicate and divide without killing them, and the cells remain able to grow if the antibiotic is removed; or (2) bactericidal, in which case the cells are killed.

An antibiotic may be bactericidal because it interacts with a cell structure so as to damage irreversibly its integrity or function, or because it binds to an essential enzyme or cell structure with such a high affinity that the binding is practically irreversible. The bacteriostatic antibiotics bind with less affinity to cell structures on which they act, so that when the antibiotic is removed from the bacterial environment the cell structure–antibiotic complex dissociates and the cell structure again becomes active.

At the molecular level, cell duplication can be divided into the following stages (Fig. 3.1):

1. transformation of the carbon and nitrogen sources of the culture medium into key products for metabolism and chemical energy;

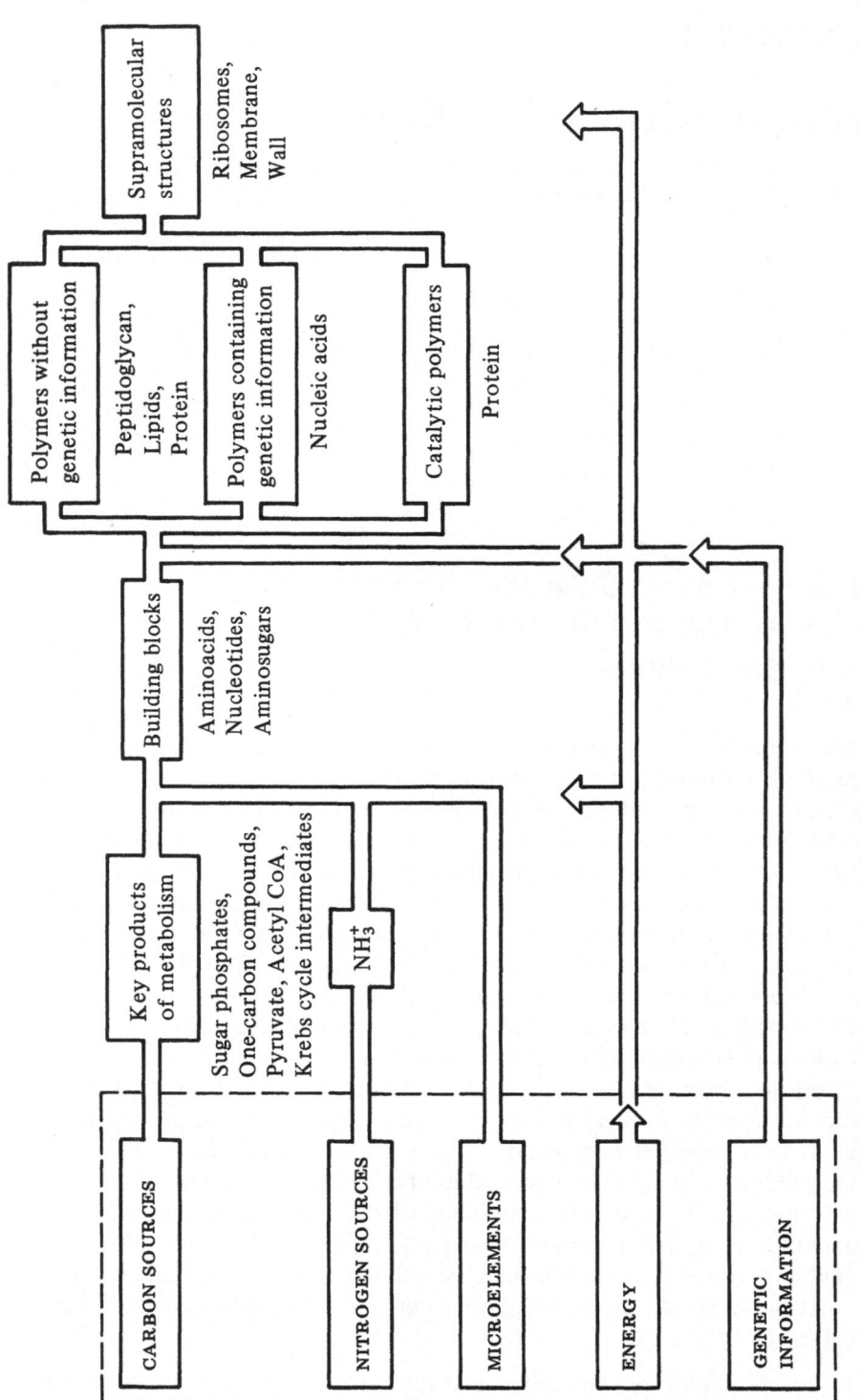

Figure 3.1. The flow of matter, energy, and information during cellular growth.

2. production of monomers (amino acids, nucleotides, glucosamines, triglycerides, etc.) and cofactors;
3. polymerization of monomers into macromolecules (proteins, nucleic acids, peptidoglycans);
4. integration of the macromolecules into supermolecular structures (ribosomes, cell walls, cytoplasmic membranes, etc.).

Processes 2–4 use energy produced during process 1. Process 3 follows a rigorous code of information that determines the order in which the monomers are polymerized.

Antibiotics can inhibit cell growth by interfering with any one of the processes listed above. The mechanism of inhibition is called the mechanism or mode of action of the antibiotic. Antibiotics can be classified on the basis of their mechanisms of action:

1. inhibitors of transformation of carbon or nitrogen sources into monomers;
2. inhibitors of polymerization or of formation of supermolecular structures;
3. inhibitors of energy metabolism;
4. substances that interfere with the flow of genetic information, with consequent formation of polymers that do not correspond to the authentic information of the cell;
5. antibiotics that interact with already formed cell membranes to alter their permeability.

A. Selectivity of Action of the Antibiotics

The science of antibiotics is essentially an applied science. It is directed at discovering and using substances that can inhibit the growth of microorganisms parasitic in man or in the higher animals without damaging the host organism. Therefore, the antibiotic must be selective in its mechanism of action: it must interact with the bacterial cell and not with the cells of the host.

1. Evolutionary Basis of Selectivity

The selectivity of antibiotic action derives from the structural and biochemical variations that have arisen during evolution without changing the essential functions of the cell.

On the evolutionary scale, one can distinguish between two principal groups of organisms: prokaryotes and eukaryotes. The prokaryotes include all the bacteria and blue algae. All other cellular organisms (yeasts, fungi, algae, protozoa, plants, and animals) are eukaryotes. The most important structural and biochemical differences are summarized in Table 3.1 and Fig. 3.2.

Table 3.1. Summary of the Essential Differences between Prokaryotes and Eukaryotes Responsible for the Selective Toxicity of Antibiotics

Structure	Eukaryotes	Prokaryotes
Nucleus	True nucleus enclosed	No nuclear membrane
Chromosomes	Several chromosomes made up of DNA and protein	One chromosome made up of DNA only
Mitotic apparatus	Presence of centriole and spindle	Absence of mitotic apparatus; role played by plasma membrane
Cytoplasmic membrane	Sterols present	Sterols absent
Cell wall	Absent (animals) When present (algae, plants and fungi) made up of simple polysaccharides	Typically a complex structure with peptidoglycan and atypical amino acids
Ribosomes	80S except for those of mitochondria and chloroplasts	70S
Respiratory systems	Located in specialized organelles, the mitochondria	Located on plasma membrane
Other cytoplasmic organelles	Present (e.g., chloroplasts)	Absent
Cellular osmotic pressure	Similar to that of environment	Higher than that of environment

2. Cellular and Molecular Bases of Prokaryote versus Eukaryote Selectivity

Why are antibiotics that effectively inhibit bacterial cells at low concentrations inactive even at much higher concentrations against animal cells? In other words, what is the cellular or molecular basis of the selectivity of antibiotic action? Antibiotics are a very heterogeneous group of substances, both in terms of structure and mechanisms of action. Therefore, one should expect that their specificities would have different bases. We can propose three major classes of mechanisms to explain the selectivity of the antibiotics: (1) differences in degrees of penetration into different organisms, (2) actions on essential structures typical of microorganisms and not present in animal cells, (3) different affinities of the antibiotic for molecules that may play the same role in different types of cell but are not identical in all cells.

a. Differences in Permeability

An antibiotic is active if it enters into the cell in a concentration sufficient to inhibit cell growth. The intracellular concentration of an antibiotic at any one moment is the resultant of its rate of penetration from the out-

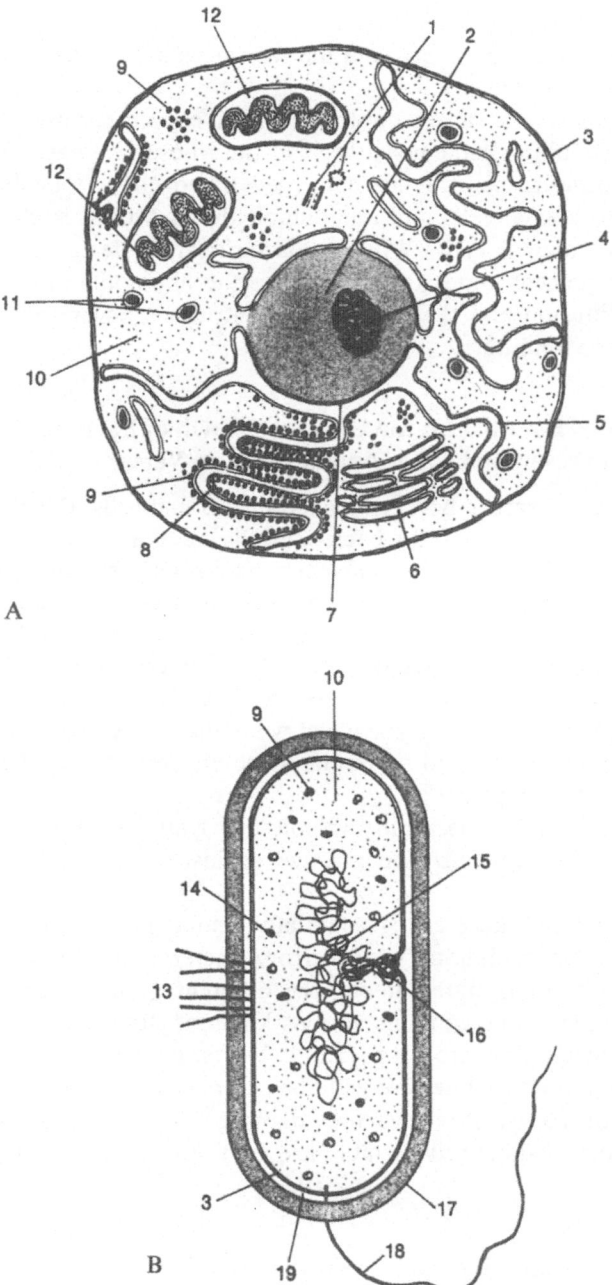

Figure 3.2. Schematic drawing of the ultrastructural organization of a generic eukaryotic cell (**A**) and of a typical bacterial cell (prokaryotic) (**B**). (1) Centriole; (2) nucleus; (3) membrane; (4) nucleolus; (5) smooth endoplasmic reticulum; (6) Golgi apparatus; (7) nuclear membrane; (8) rough endoplasmic reticulum (9) ribosome; (10) cytoplasm; (11) lysosome; (12) mitochondrion; (13) pili; (14) granular inclusion; (15) chromosome; (16) mesosome; (17) capsule; (18) flagellum; (19) wall.

side to the inside of the cell and the rate at which it becomes diluted as a result of cell division.

Tetracyclines are an example of selective permeability. These antibiotics, which inhibit protein synthesis in the ribosomes, have approximately equal activities in cell-free systems (cell homogenates) made from either bacterial or animal cells, but they inhibit the growth of bacterial cells and not that of animal cells. Whereas bacteria are able to take up actively inhibitory concentrations of tetracycline, the antibiotic enters animal cells only by diffusion and the internal concentrations never reach levels high enough to be inhibitory.

b. Structures Essential for the Life and Growth of Bacteria but Not Present in Higher Animals or Vice Versa

An essential difference between bacterial cells and cells of higher organisms (especially animals) is that bacterial cells have a rigid cell wall outside the cellular membrane that prevents the bacterium from bursting as a consequence of its high internal osmotic pressure (see section III, this chapter). Elimination or weakening of the cell wall enables it to burst and kills the bacterium. (The capacity of bacterial forms to exist under exceptional conditions without a cell wall is discussed in section III, this chapter.) Yeasts and fungi have cell walls that are completely different in chemical composition and molecular structure from those of bacteria (see section III, this chapter).

Some antibiotics specifically inhibit the synthesis of the bacterial cell wall. They will obviously be inactive against yeast, fungus, and animal cells.

The cell membranes are substantially similar in structure and function throughout the evolutionary scale from bacteria to mammals. However, there are small variations among the different groups. For example, the eukaryote membrane contains various kinds of sterols, which vary in different organisms, that are not present in bacterial membranes.

Certain antibiotics interfere with the cell membrane sterols of eukaryotes to an extent that causes irreversible damage. Since bacteria do not possess these sterols in their cell membranes, they are insensitive to those antibiotics.

c. Different Affinities of the Antibiotic for Structures Common to Prokaryotes and Eukaryotes

The principal metabolic pathways and structures (e.g., ribosomes) in the bacteria are also present in yeasts, fungus, and animal cells. However, enzymes that catalyze the same reactions can have slightly different structures. These modifications, which have been acquired during evolution, can change the affinity of binding with a given antibiotic. For this reason, much greater antibiotic concentrations may be required to inhibit the enzyme in the animal cell than in the bacterial cell, or vice versa.

This is the most common type of selectivity and applies to almost all antibiotics that inhibit protein synthesis enzymes, RNA polymerases, and dihydrofolate reductases.

3. Cellular and Molecular Bases for Differences in Antibacterial Spectrum

Antibiotics show differences not only when comparing activities against prokaryotes and eukaryotes, but they also show substantial differences in activity against different species within the prokaryote group. In other words, they have different antibacterial spectra. The cellular and molecular bases for the differences in the antimicrobial spectra of different antibiotics can be grouped into three classes: (1) differences in permeability, (2) either the presence or absence of inactivating enzymes, (3) either the presence or absence of a molecular target.

a. Differences in Permeability

Most antibiotics, unlike molecules that are recognized by the cell as nutrients (sugars, amino acids, etc.), enter the cell by diffusion at rates proportional to concentration gradients between the outside and the inside of the cell and dependent on the physiochemical nature of the barrier. The barrier consists of the cellular membrane and the cell wall. The cellular membranes of different bacteria have very similar structures and compositions, whereas the cell walls differ substantially from one bacterial group to another.

The cell wall has been shown to play an essential role in determination of the selectivity of the actions of different antibiotics. The cell walls of gram-negative bacteria offer a greater "resistance" to diffusion than the less complicated cell walls of Gram-positive species. In fact, the latter group are generally more sensitive to antibiotics than Gram-negative bacteria. The outer membrane of Gram-negative bacteria has been implicated as the major barrier to diffusion.

It is possible to remove the outer membrane of a Gram-negative cell (e.g., by manipulation of the ion concentration) while leaving the peptidoglycan layer and the cellular membrane unaltered. Bacteria so treated become much more sensitive to many antibiotics. An antibiotic, in entering the cell, must penetrate the lipophilic external membrane, the hydrophilic peptidoglycan, the lipophilic cell membrane, and, finally, the cytoplasm. A substantial rate of diffusion can occur only if the antibiotic possesses an "appropriate" degree of lipophilicity. If it is too hydrophilic, it does not penetrate the external membrane; if it is too lipophilic, it tends to concentrate in the lipophilic layer of the wall and does not diffuse into the more hydrophilic cytoplasm.

If all antibiotics entered the cell by diffusion, it would be necessary to exclude the existence of water-soluble antibiotics, which would not be able to cross the lipid layer. Such antibiotics, however, do exist, e.g., the aminoglycosides. They are presumably transported into the cell by an "energy-consuming active mechanism" that uses specific carriers. Although such carriers have not been identified, the presence of active transport is strongly suggested by the finding that conditions that deplete the supply of cellular energy reduce the transport of these antibiotics and consequently induce resistance.

b. Presence of Inactivating Enzymes

Several bacterial species possess the capacity to produce enzymes that inactivate some antibiotics by modifying their chemical structure. This property is particularly evident in species of *Pseudomonas* genus, which for this reason (in addition to their impermeability) are not susceptible to the vast majority of known antibiotics. The inactivating enzymes, which are discussed in Chapter 4, include acetylating enzymes, phosphorylating enzymes, and peptide-splitting enzymes (peptidases).

c. Absence of a Molecular Target

Although most antibacterial antibiotics act on structures present in all bacterial species there are some example of inactivity due to lack of an appropriate target. The best known representatives are mycoplasmas, bacteria devoid of a cell wall. Penicillins and the other β-lactam antibiotics, because they act by interfering with cell wall synthesis, have no effect on these bacteria.

Another example is bicyclomycin, which is inactive on Gram-positive bacteria since it acts by disrupting the cell wall outer membrane, a structure typical of Gram-negative bacteria.

II. Methods for Study of the Mechanism of Action of Antibiotics

To oversimplify considerably, we can say that the means to uncover the mechanism of action of a new antibiotic is to study its activity: (1) in the intact cell, (2) in a partially purified cell-free system, and (3) in a purified enzyme system.

A. Activity in the Intact Cell

The most generally used procedure is to add the antibiotic to growing culture of sensitive bacteria and to observe the effects on synthesis of bacterial macromolecules, i.e., DNA, RNA, proteins, and peptidoglycan. This synthesis can be easily followed by adding radioactively labeled specific precursors to the culture medium.

Labeled thymine is used to follow the synthesis of DNA, labeled uracil for RNA, labeled phenylalanine (or other amino acid) for protein, and labeled acetylglucosamine for peptidoglycan. At regular intervals one measures the amount of radioactivity incorporated into the corresponding macromolecule by the bacteria, which is an index of the amount of synthesis. Figure 3.14 shows the synthesis of the macromolecules in a growing bacterial culture treated with chloramphenicol. The primary effect is inhibition of protein synthesis, as shown by the rapid arrest in phenylalanine incorporation.

Naturally, when synthesis of a macromolecule essential for cell growth is stopped, it has effects on all the other cellular functions, with a consequent arrest of the synthesis of other types of macromolecule. Therefore, to establish what the primary effect is, one must observe the times at which the various events can be detected. The primary effect is usually that seen earliest, but there can be simultaneous arrest of more than one or even of all the macromolecular syntheses; for example, when the antibiotic acts on the respiratory function or on the integrity of the cell membrane.

B. Activity in Cell-Free Polymerizing Systems

After one has determined the primary effect of the antibiotics, which is to say which macromolecular synthesis is inhibited first, one must ascertain whether the inhibition is due to interference with (1) synthesis of precursors or with their activation, (2) the enzymes or structures involved in polymerization, or (3) the information system that determines the order in which the precursors are lined up into the polymer.

This type of study is usually carried out with enzyme systems obtained from partially purified cell homogenates that are able to polymerize the monomers into macromolecules. Cell-free systems for in vitro synthesis of proteins, nucleic acids, and peptidoglycan can be prepared. If the antibiotic inhibits the synthesis of the same macromolecule in vitro, as was inhibited in the growing cell, then this indicates that it acts on the polymerization process, or possibly in some cases on the information molecules. If it does not, one should look for its effect on synthesis of precursors or their activation.

To return to the example of chloramphenicol, which we have already said inhibits protein synthesis in the intact cell, one must determine its

capacity to inhibit the activity of an enzyme system for amino acid polymerization, in a cell extract that contains among other things ribosomes,
RNA, and several enzymes and factors. Chloramphenicol inhibits the activity of this system, and one can therefore deduce that the arrest in protein
synthesis observed in the growing cells is due to a direct interference with
the amino acid polymerization process.

C. Activity in Purified Enzyme Systems

The polymerizing systems mentioned above usually contain many components, any one of which might be the target of the antibiotic's effect.
Within certain limits one can separate the principal enzyme reactions of the
system and determine exactly where the antibiotic acts.

To continue with the discussion of chloramphenicol as an example,
looking at the various components of protein synthesis in vitro, one can
see that this antibiotic inhibits protein synthesis by interfering with the
ribosome, more precisely, with the 50S subunit, and to be still more precise, with one protein of the 50S subunit, peptidyltransferase.

These methods use biochemical and molecular biology techniques. Also
of great importance are genetic techniques. Often the definitive proof that
the mechanism of action of an antibiotic is inhibition of a certain enzyme
can be obtained by isolating from a population of bacteria a mutant that
is resistant to the antibiotic and showing that the antibiotic does not inhibit in vitro the enzyme extracted from the resistant bacteria as it had
inhibited the enzyme extracted from the sensitive bacteria.

Genetic techniques can also be used to obtain preliminary indications
about the mechanism of action of a new antibiotic. For example, if there
is cross-resistance between the new antibiotic and another antibiotic whose
mechanism of action is known, it is probable that the target of the first
will be the same or very similar to that of the second. If there is no cross-
resistance, another site of action is implied. This obviously makes it easier
to know what biochemical experiments must be done to confirm the
hypothesis.

III. Inhibitors of Cell Wall Synthesis

Before discussing the mechanisms of action of inhibitors of cell wall synthesis, we shall briefly describe the chemical structure of the cell wall of
different microorganisms and the principal biosynthetic pathways.

A. Structure and Architecture of the Cell Wall

The cell wall is a rigid structure that encloses the microbial cell, determines its shape, and protects it from bursting as a consequence of the high internal osmotic pressure. In the prokaryotes, the fundamental structure consists of a complex three-dimensional network comprised of peptidoglycan, a macromolecule (mucopeptide or glycopeptide or murein), and other polymers (polysaccharides, lipoproteins, etc.) that vary in different types of bacteria. The peptidoglycans consist of long polysaccharide filaments formed from two monomers, acetylmuramic acid (M) and acetylglucosamine (G), arranged alternately and connected by a $\beta 1 \rightarrow 4$ glycoside linkage. The filaments are interconnected by four amino acid peptides, in which the third amino acid is variable, that branch off from the acetylmuramic unit to form the three-dimensional structure (Figs. 3.3 and 3.4).

The manner in which the peptide chains are interconnected differs among bacterial species. In some cases they are connected directly, and in others amino acid chains are used for the linkage.

In addition to the three-dimensional network of peptidoglycan, there are other structures in the cell wall that differ from organism to organism, but that are basically of two types: those found in the Gram-positive bacteria and those found in the Gram-negative bacteria.

1. Gram-Positive Cell Walls

These consist of a layer that appears uniform when viewed under the electron microscope and is made up of peptidoglycan, protein, and teichoic acids. These last are widely dispersed and present in considerable quantity. In all Gram-positive bacteria one also firds a "membrane teichoic acid," so called because it is situated between the wall and the cytoplasmic membrane. It consists of a chain of alternating glycerols and phosphates. The teichoic acids of the wall are bound covalently to the peptidoglycan. They can have a structure similar to that described above or may contain ribitol instead of glycerol. In some rare forms of teichoic acid, the basic polymer contains a sugar such as glucose or N-acetylglucosamine in addition to the polyalcohol (glycerol).

2. Gram-Negative Cell Wall

The cell wall is thinner in Gram-negative than in Gram-positive bacteria but is more complex in the former. It is typically composed of several layers. The innermost layer is peptidoglycan, which is separated from the cytoplasmic membrane by a periplasmic space in which several enzymatic activities are found. On the external surface of the peptidoglycan layer lies the outer membrane. This can be somewhat schematically described as a double layer of lipopolysaccharide giving rise to a structure similar

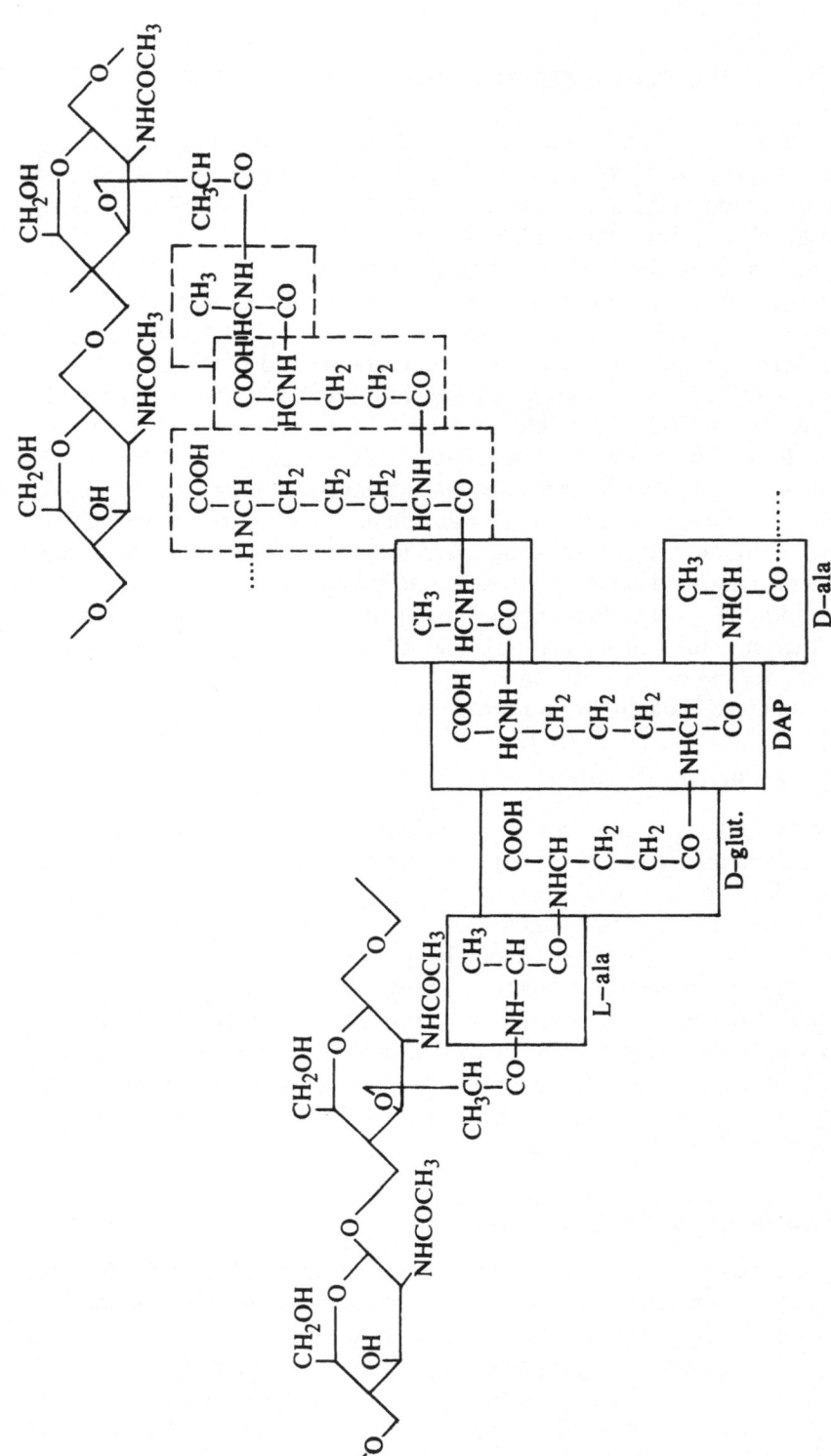

Figure 3.3. Chemical structure of peptidoglycan (type 1).

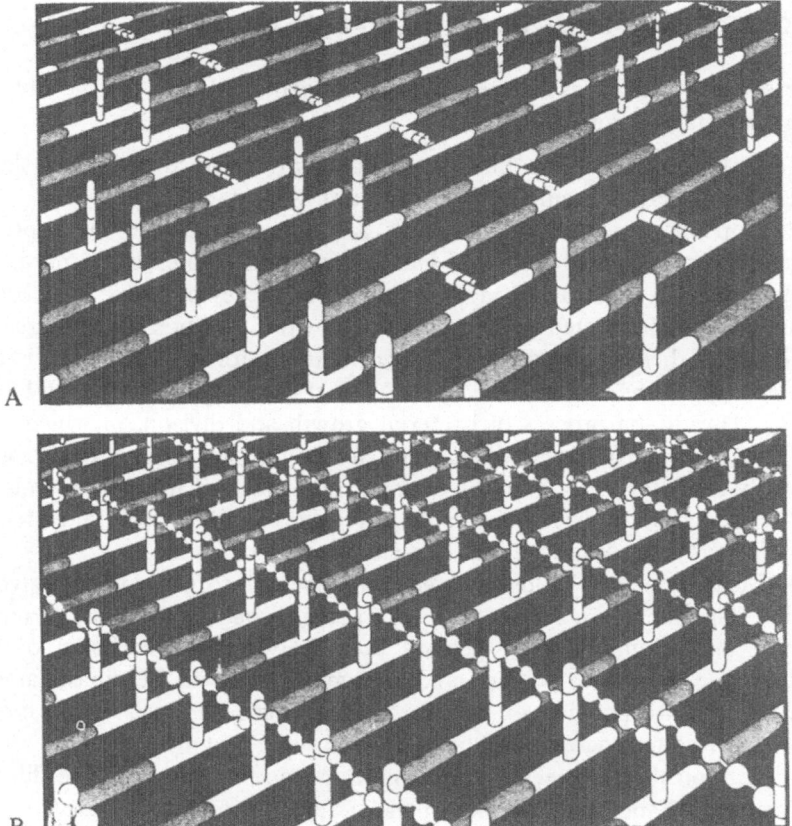

Figure 3.4. Schematic representation of a layer of the peptidoglycan network in *E. coli* (**A**) and in *S. aureus* (**B**). The long parallel horizontal chains are made up of N-acetylglucosamine (gray sections) and muramic acid (white sections). Tetrapeptides are attached to the muramic acid residues (segmented white pieces). In *E. coli* only some tetrapeptides are interconnected, with the bond always between the third amino acid of one tetrapeptide and the fourth aminoacid of the adjacent tetrapeptide. In *S. aureus* the interpeptide bonds are indirect, through a pentaglycine bridge (*white dots*). (From "Il meccanismo d'azione degli antibiotici," in M. Vergnano and D. Sassella (eds.), *Rassegna Medica*. Milan, Italy: Gruppo Lepetit, Spa., 1973, p. 33.)

to the cytoplasmic membrane. Proteins cross the layers, some of which have permeability functions. A complex lipopolysaccharide layer covers the external surface of the outer membrane.

3. Fungal Cell Walls

These are made up of polysaccharide fibers interlaced within the amorphous polysaccharide. The fibrils are usually of chitin (poly-N-acetylglucosamine, $\beta 1 \rightarrow 4$).

B. Biosynthesis of Peptidoglycan

The series of reactions in the synthesis of peptidoglycan can be divided into three stages (Figs. 3.5 and 3.6):

1. Production of basic units, starting from intermediary metabolites. These are UDP-glucosamine and UDP-muramylpentapeptide.

2. Condensation of the basic units and their transfer to the peptidoglycan structure being built up. The activated molecule, UDP-muramylpentapeptide, is attached to a lipid carrier in the cytoplasmic membrane (this is a phosphorylated isoprenoid alcohol with 55 carbon atoms), freeing UDP, and it then reacts with UDP-glucosamine to form a glycoside linkage between C-1 of the muramic acid and C-4 of the acetylglucosamine (B1 → 4 bond). During the process of bacterial growth and divison, in which new peptidoglycan must be made, enzymes on the internal surface of the wall partially hydrolyze the three-dimensional network and make free ends of the peptidoglycan molecule available to receive the dimer (muramylpentapeptide–acetylglucosamine) and thus elongate the chain.

3. Formation of bonds between the newly formed peptidoglycan chains. The interchain connections take place through loss of the terminal alanine and formation of a peptide bond between the carboxyl of the D-alanine in the fourth position and the amine group of the dibasic amino acid in the third position on another chain. This reaction is called cross-linking. In some bacteria the bond is not direct but by way of a small peptide chain that connects the D-alanine in one filament with the dibasic amino acid of another. The loss of the D-alanine provides an explanation of why the structure of the completed peptidoglycan contains tetrapeptides instead of the pentapeptide precursor (acetylmuramylpentapeptide).

Although the basic characteristics of peptidoglycan synthesis are those described above, there seem to be several types of peptidoglycan maturation. At the very least, one can distinguish two types: one associated with cell elongation and one involved in septum formation. However, other types may exist, as demonstrated by the isolation of mutants that show morphological deformities when they have lost one type of synthesis capacity.

C. Inhibitors of Cell Wall Synthesis

The inhibitors of cell wall synthesis can be divided into two classes (Fig. 3.7): (1) inhibitors of peptidoglycan synthesis and (2) inhibitors of synthesis or assembly of other components of the wall. The large majority of known cell wall synthesis inhibitors belong to the former class. Bicyclomycin is the only antibiotic of the latter class whose mechanism of action has been sufficiently elucidated. It has been shown to inhibit the synthesis (or assembly) of lipoprotein of the outer membrane of Gram-negative

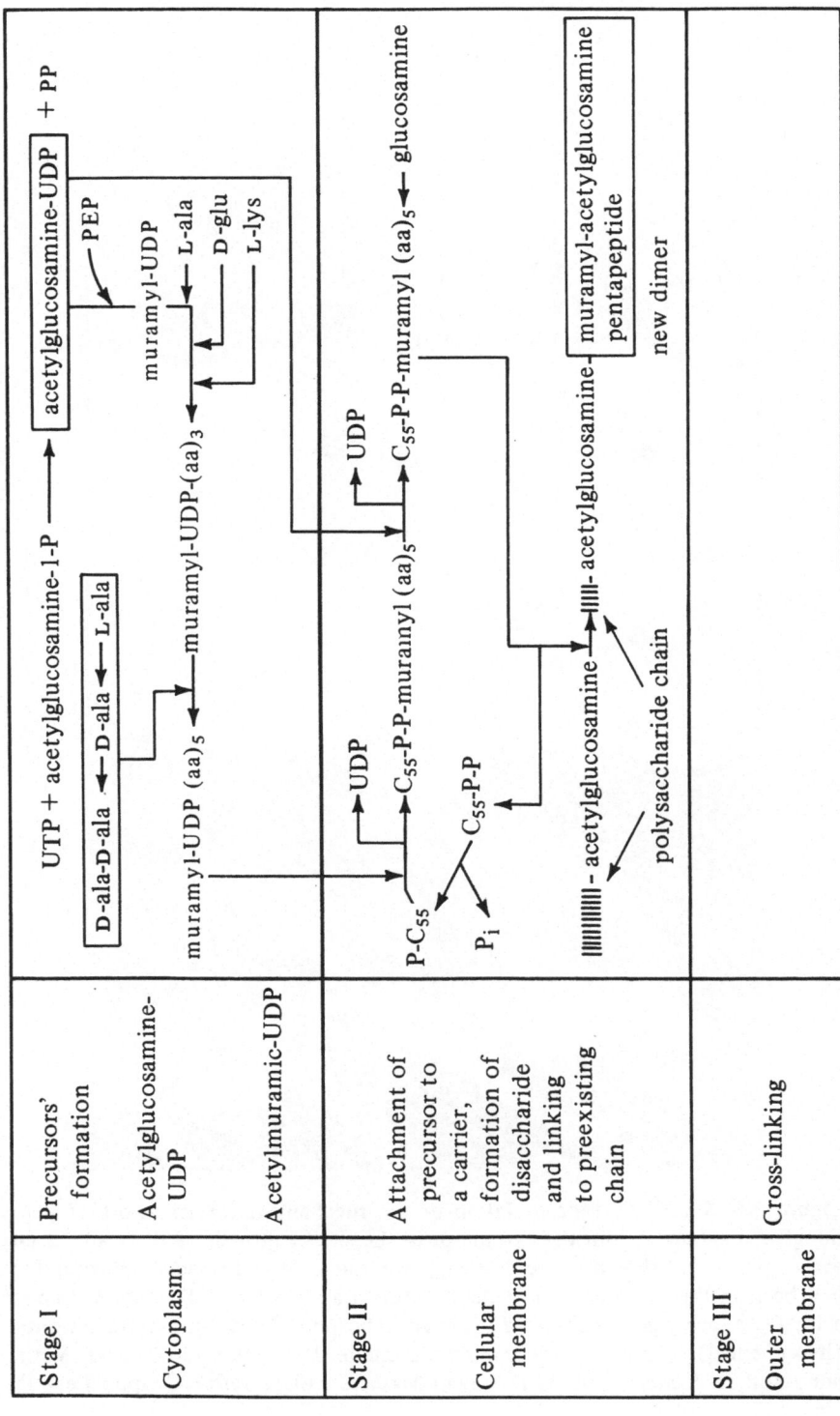

Figure 3.5. Biosynthesis of peptidoglycan.

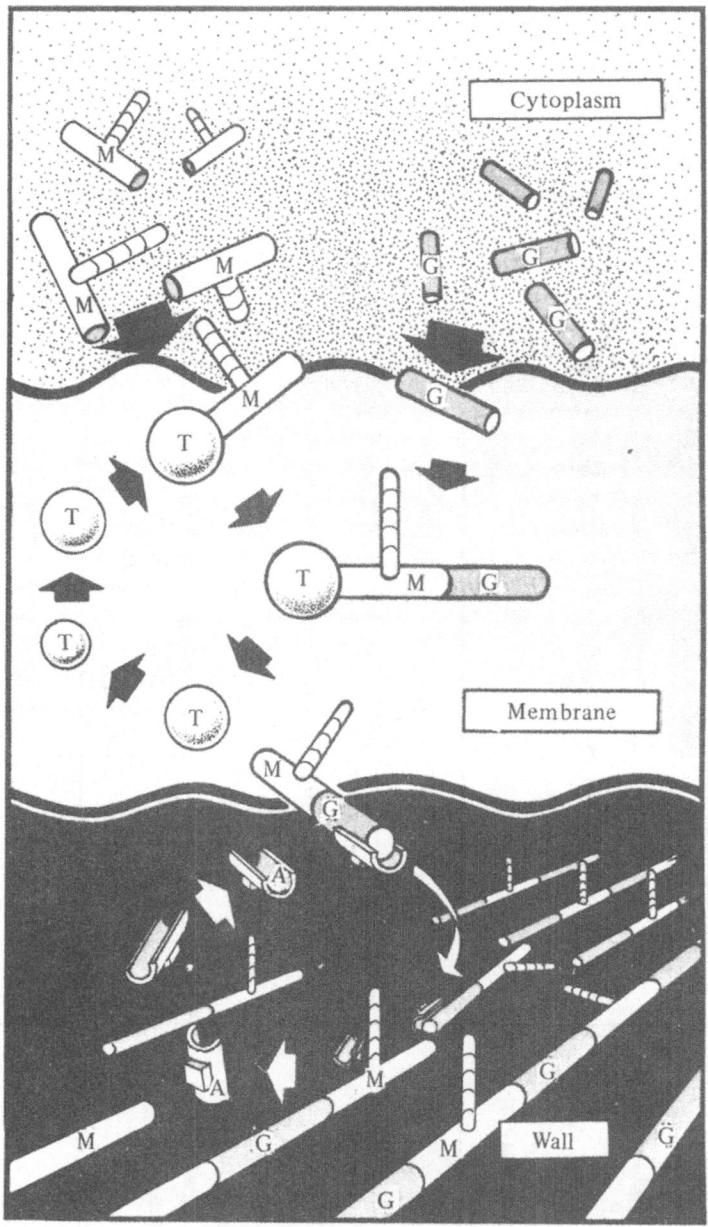

Figure 3.6. Schematic representation of the mechanism for transport of peptidoglycan subunits from the cytoplasm to the site of growth of the wall. In an earlier phase, not shown in the drawing, muramic acid and N-acetylglucosamine had been in the cytoplasm in nucleotide form, attached to UDP. M, Muramic acid; G, N-acetylglucosamine; T, carrier (phosphorylated lipid); A, acceptor (hypothetical). (From "Il meccanismo d'azione degli antibiotic," in M. Vergnano and D. Sassella (eds.), *Rassegna Medica.* Milan, Italy: Gruppo Lepetit, Spa., 1973, p. 34.)

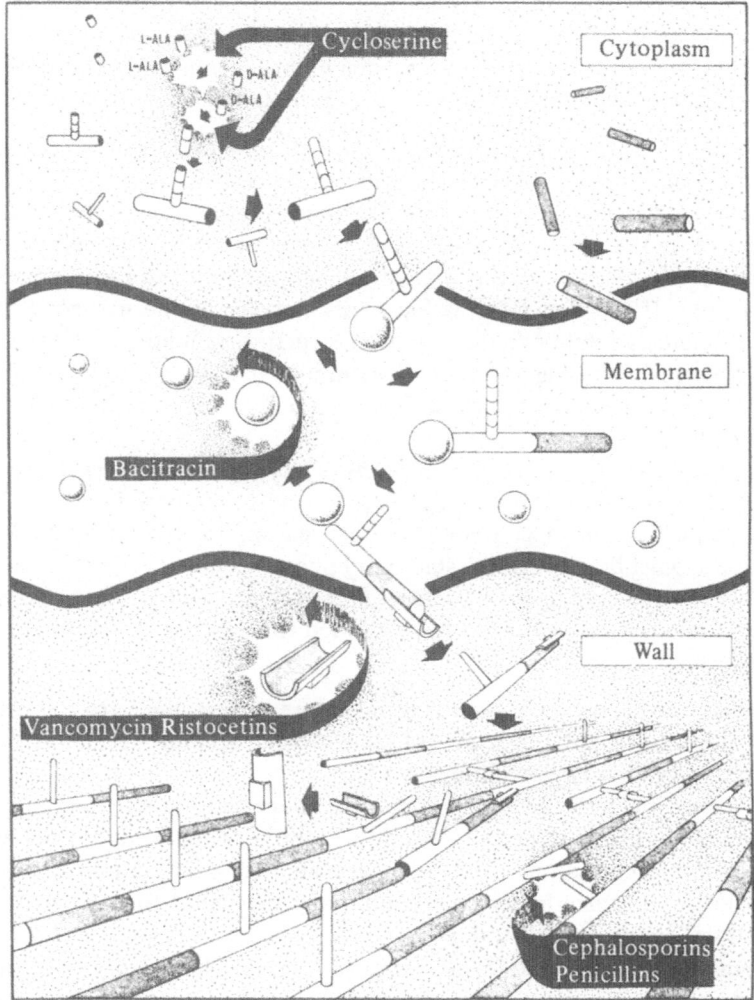

Figure 3.7. Schematic representation of the site of the metabolic effects of some inhibitors of cell wall synthesis. (From "Il meccanismo d'azione degli antibiotic," in M. Vergnano and D. Sassella (eds.), *Rassegna Medica*. Milan, Italy: Gruppo Lepetit, Spa., 1973, p. 46.)

bacteria. Obviously it is active only on Gram-negatives, not on Gram-positives.

The inhibitors of peptidoglycan synthesis can be conveniently classified into three groups according to the process with which they interfere:

1. inhibitors of the formation of the building blocks, e.g., cycloserine, phosphomycin;
2. inhibitors of formation of the dimer and inhibitors of its transfer to the

growing peptidoglycan chains, e.g., vancomycin, ristocetin, bacitracin, flavomycin;
3. inhibitors of cross-linking, e.g., penicillins, cephalosporins, cephamycins, thienamycin (two-ring β-lactams), monobactams.

Regardless of their precise sites of action, all known cell wall synthesis inhibitors have several characteristics in common.

They are bactericidal antibiotics. During bacterial growth, lytic enzymes are active on the inner site of cell wall. They break open the peptidoglycan chains to give rise to "free ends" into which the newly formed dimers are transferred. If the dimer formation or its transfer to the nascent peptidoglycan chain is inhibited, the cell wall structure remains loose and, as a consequence of the high osmotic pressure of the cytoplasm, the cell wall breaks open and the cytoplasm flows out (Fig. 3.8) (see also β-lactam antibiotics).

They are inactive against resting cells. The lytic enzymes associated with cell wall synthesis are active only when the cells are growing but not when they are resting. As a consequence, the cell wall synthesis inhibitors are inactive against bacteria in the stationary phase.

They are inactive against bacteria that lack cell walls. The existence of bacteria without cell walls has been alluded to in section I. There are three types of such bacteria:

1. *Mycoplasms,* bacteria occurring in nature, where they are parasites of animals and plants; they do not have a cell wall and are unable to synthesize it under any conditions.
2. *L-forms,* bacteria generally obtained under laboratory conditions but sometimes found in nature (e.g., in infections of the urinary tract, where hypertonic conditions exist) that are able to grow and to multiply if maintained under hypertonic conditions; they tend to regenerate their cell walls unless a cell wall synthesis inhibitor is present.
3. *Protoplasts,* bacterial forms produced in the laboratory, as are the L-forms, but which, unlike the latter, are unable to multiply. Myco-

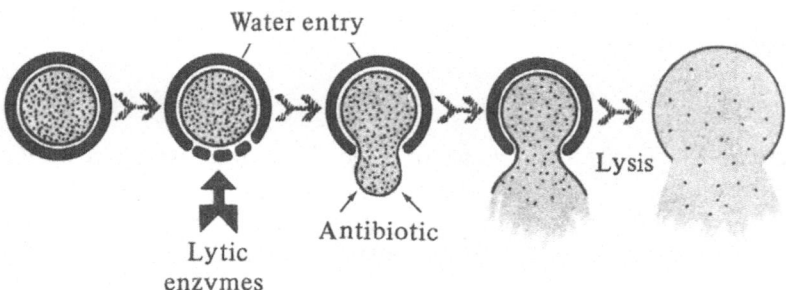

Figure 3.8. Lytic effect on growing cells of antibiotics that inhibit cell wall synthesis.

plasms, L-forms, and protoplasts are insensitive to cell wall synthesis inhibitors.

D. Examples of Cell Wall Synthesis Inhibitors

We now briefly describe the modes of action of the best known antibiotic inhibitors of peptidoglycan synthesis. (For other properties of these antibiotics, see Chapter 5.)

1. Phosphomycin

Phosphomycin (formerly known as phosphonomycin) is a small molecule that specifically inhibits the reaction of ULP-N-acetylglucosamine with phosphoenolpyruvate to form UDP-N-acetylmuramic acid. This reaction is catalyzed by the enzyme pyruvyl transferase. Phosphomycin binds covalently to this enzyme to cause irreversible inhibition.

2. Cycloserine

Cycloserine is an antibiotic that is particularly active against mycobacteria. Its antibiotic effects can be antagonized by addition of D-alanine to the culture medium. This suggests that cycloserine acts at the terminal portion of the pentapeptide of muramic acid. It was then demonstrated in vitro that it competitively inhibits (1) the enzyme that converts L-alanine into its stereoisomer D-alanine (alanine racemase) and (2) the enzyme that catalyzes the formation of the peptide bond between two molecules of D-alanine, with formation of D-alanyl-D-alanine (D-alanyl-D-alanine synthetase).

Cycloserine is structurally similar to one of the possible conformations of D-alanine and is therefore erroneously recognized as D-alanine by the enzyme and binds to it, inhibiting its catalytic function.

3. Bacitracin A

Bacitracin A is a polypeptide antibiotic isolated from broth cultures of *Bacillus subtilis*. It blocks the phosphorylase that liberates one of the two terminal phosphates from the lipid carrier, which can then no longer function as an acceptor of muramylpentapeptide. However, bacitracin is not very specific and it also inhibits other phosphorylase reactions. It is this lack of specificity that most likely is the cause of its high toxicity.

4. Vancomycin and Ristocetin

These are complex glycopeptide molecules containing aromatic amino acid derivatives and sugar derivatives. They inhibit the transfer of muramylpentapeptide–acetylglucosamine from the lipid carrier to the peptidoglycan chain being formed.

5. β-Lactam Antibiotics

This group of antibiotics includes the penicillins, the cephalosporins, the cephamycins, and the newer β-lactams such as thienamycin, PS-5, and clavulanic acid. All have similar but not identical mechanisms of action: They prevent peptidoglycan maturation by inhibiting the cross-linking reaction. Two enzymes are involved, the transpeptidase that removes the terminal D-alanine from N-acetyl-muramylpentapeptide and links the remaining tetrapeptide to another part of the peptidoglycan structure and the carboxypeptidase that removes the terminal D-alanine without catalyzing cross-linkage. The activity against the transpeptidase seems to be the most important (Fig. 3.9).

In the growing bacterium, several different proteins of the inner membrane have been identified that are able to specifically bind β-lactams, penicillin binding proteins (PBP). However, different β-lactams have different degrees of affinity for different binding proteins. The extreme case is that of mecillinam, which binds to only one. The different binding proteins may be responsible for the different types of peptidoglycan maturation. As a consequence of the differential inhibition of the various types of peptidoglycan maturation, the different β-lactams cause a variety of morphological effects. For example, benzylpenicillin has a lytic action, and

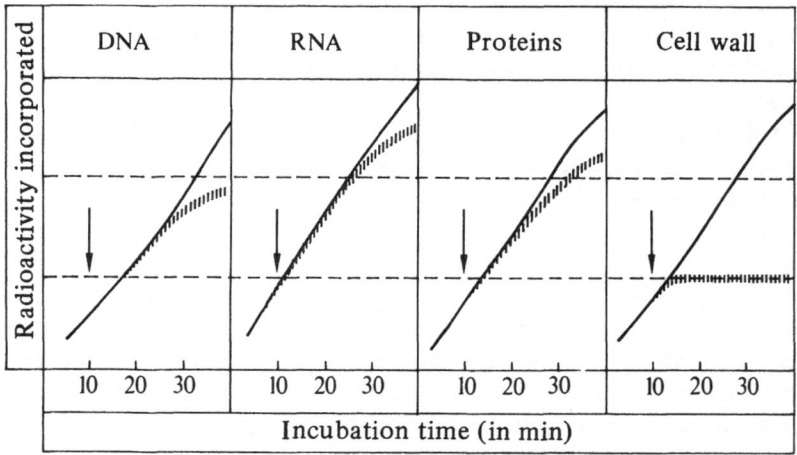

Figure 3.9. Effect of penicillin on incorporation of radioactive precursors into DNA, RNA, proteins, and cell walls of growing bacteria. The *solid line* indicates incorporation into a control culture, the *dashed line* incorporation into a culture to which antibiotic was added at the time indicated by the *arrows*.

cephalexin induces the formation of elongated forms of *E. coli,* in which septum formation is specifically inhibited while elongation remains unaffected. Mecillinam, on the other hand, induces the formation of round forms, in which cell elongation, but not septum formation, is inhibited.

The various PBPs may be different forms of transpeptidases and carboxypeptidases.

How do β-lactam antibiotics lyse susceptible microorganisms? A few years ago, the mechanism of action of β-lactams seemed fully understood. The key features of the proposed model of action mechanism were: (1) a critical biochemical target in cell wall synthesis (one or more enzymes responsible for the transpeptide cross-linking of the newly formed peptidoglycan chains) is inhibited by β-lactams because of their structural similarity to D-alanyl-D-alanine, a substrate of the transpeptidase reaction; (2) the uncrossed-linked building blocks are incorporated into the growing peptidoglycan chains, generating a "weak" cell wall; (3) the uninhibited accumulation of cytoplasmic material exerts increasing pressure on the "weak" cell wall, eventually resulting in wall rupture and cell lysis. According to this model, cell lysis would be a "direct" consequence of "unbalanced" growth, i.e., inhibited cell wall synthesis and uninhibited cell mass growth.

Recent experimental results, however, are suggesting a more complex situation. The major experimental data conflicting with the above described model include the following: (1) several bactericidal β-lactams do not inhibit cross-linking; (2) in some bacteria, the poorly cross-linked peptidoglycan elements are not incorporated into growing chains and thus the presumed "weakening" of cell walls does not exist; (3) mutants have been isolated in which penicillin treatment results in growth inhibition but not cell lysis (penicillin-tolerant strains). These strains either lack a peptidoglycan hydrolytic enzyme(s) or have an excess of an inhibitor of such enzyme(s).

These findings led to the proposal of new model of mode of action of β-lactams. According to this model the irreversible effects of β-lactams (cell lysis) would be only "indirectly" related to the antibiotic's interference with their primary targets (PBPs). The model operates as follows: (1) β-lactams inhibit the activity of one or more enzymes in peptidoglycan synthesis (PBPs) resulting in inhibition of bacterial growth; (2) a "signal" (of yet unknown biochemical nature) is generated that triggers a release of teichoic acids into the medium; (3) the loss of teichoic acids activates one or more peptidoglycan hydrolyzing activities; (4) that sever covalent bonds of the cell wall, thus exposing the cell membrane; and (5) osmotic lysis of cell will ensue eventually.

IV. Inhibitors of Transcription and Replication of Genetic Material

A. Replication and Transcription of Genetic Information

The synthesis of nucleic acid can be divided into two phases:

1. synthesis of precursors (nucleotides and deoxynucleotides) from intermediate molecules of cell metabolism;
2. enzymatic polymerization of nucleotides to form a macromolecule whose sequence is determined by the sequence of the bases in the DNA template.

Many different enzymes are involved in the replication of DNA. A simplified scheme of DNA replication is shown in Fig. 3.10. The following steps can be identified:

1. The strands of the double helix are unfolded by the action of two enzymatic proteins—a DNA-dependent ATPase and the enzyme gyrase—and are kept unfolded by binding to a specific protein, called Alberts protein or protein 32.
2. A specific RNA polymerase synthesizes short chains of RNA on each strand, presumably corresponding with specific initiator sites.
3. The replicative DNA polymerase (polymerase III of *E. coli*) synthesizes small DNA pieces (called Okazaki fragments) starting from the 3'OH of RNA, which is used as primer.
4. An RNase (in *E. coli* it is the DNA polymerase I, which, in addition to a polymerizing activity used for the repair synthesis of DNA also has nucleotidase activity) degrades the RNA fragments. DNA polymerase I completes the synthesis of the Okazaki fragments.
5. A ligase joins the adjacent fragments together.

The process of RNA synthesis has been quite thoroughly studied. RNA polymerase is an enzyme made up of five proteins, α, α', β, β', and σ. The last (σ) is essential for recognition of the initiation signal for transcription along the DNA. The functions of RNA polymerase are:

1. to form a complex with the DNA and separate the strands;
2. to incorporate the first nucleotide in the correct position on the DNA (initiation);
3. to incorporate the second nucleotide and form the first phosphodiester bond between the first and second nucleotides; to move along the DNA and continue to incorporate and form intranucleotide bonds (elongation);
4. to terminate the process when a particular sequence on the DNA has been reached (termination).

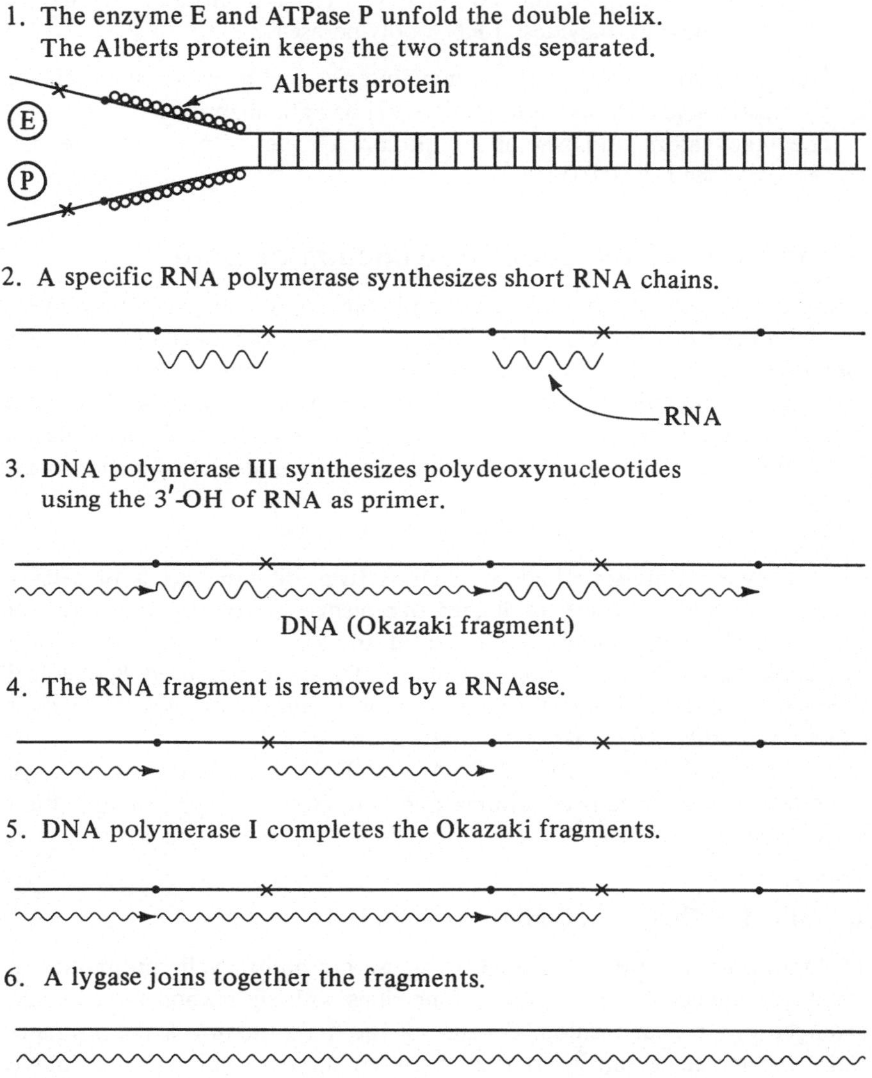
1. The enzyme E and ATPase P unfold the double helix.
The Alberts protein keeps the two strands separated.
Alberts protein
E
P
2. A specific RNA polymerase synthesizes short RNA chains.
RNA
3. DNA polymerase III synthesizes polydeoxynucleotides using the 3′-OH of RNA as primer.
DNA (Okazaki fragment)
4. The RNA fragment is removed by a RNAase.
5. DNA polymerase I completes the Okazaki fragments.
6. A lygase joins together the fragments.

Figure 3.10. Schematic model of DNA replication in *E. coli*.

B. Inhibitors of Replication and of Transcription

The inhibitors of replication and transcription can be divided into two groups:

1. inhibitors of synthesis of precursors;
2. inhibitors of polymerization. These can be divided further into (a) in-

hibitors of the template function of DNA, (b) inhibitors of the enzyme (DNA-replicating enzymes or RNA polymerase).

The precursor analogs, which may interfere with either synthesis or polymerization, are discussed in section VII as antimetabolites.

The inactivators of DNA and inhibitors of the polymerases are discussed here (see Fig. 3.10).

1. Inhibitors of the Template Function of DNA

Many substances inhibit replication and transcription because they interfere with the template function of DNA. These substances can be classified into two groups: (1) those that directly block the template function by forming a nonfunctional complex with the DNA and (2) those that cause changes in the structure of the DNA (broken strands, scission of bases, formation of covalent bonds between the two strands) so that it can no longer serve as a template.

These antibiotics have the following characteristics:

1. They bind indiscriminately to DNA from different kinds of cells—bacterial, fungal, or those of higher organisms. Therefore, they are not specific in effect and inhibit the growth of any kind of cell into which they can enter and as a result are very toxic. They are not used in medical practice as antiinfective agents but usually as antitumor agents, as their effects are particularly striking in rapidly growing cells.

2. They inhibit synthesis of RNA and DNA simultaneously, though sometimes under certain conditions one can show one type of inhibition predominating.

a. Antibiotics That Form Complexes With DNA

a1. Daunomycin and Adriamycin   These are anthracyclic antibiotics in which the molecule is made up of four rings with six carbon atoms each, arranged in a planar fashion. Because of the flat structure of the aromatic rings, the molecule can insert itself between the base pairs of the double helix of the DNA, giving a greater stability to the structure and thus inhibiting the replication of the DNA, for which the strands must come apart. The inhibition of the synthesis of RNA probably results from the fact that the anthracyclic molecule intercalates between the bases and inhibits the attachment of RNA polymerase to the DNA. Substances that behave like daunomycin are called intercalating substances. Daunomycin and its analog adriamycin are not used against infection, but only against tumors.

a2. Actinomycin D   This is one of the first antibiotics isolated and is very important historically, but it is too toxic to be used except for treatment

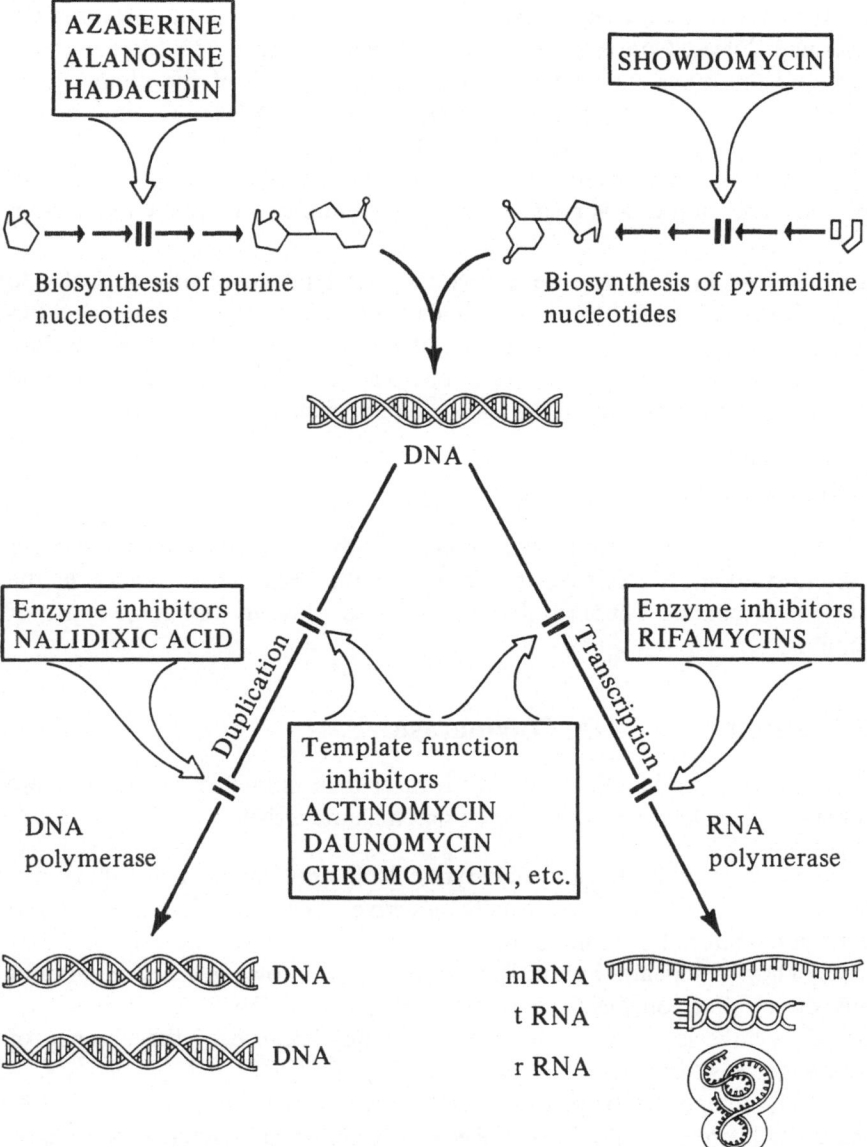

Figure 3.11. Summary diagram of the mechanisms of action of antibiotics that inhibit nucleic acid synthesis. (Modified from M. Vergnano and D. Sassella (eds.), *Rassegna Medica*. Milan, Italy: Gruppo Lepetit Spa., 1973, p. 117.)

of tumors. It forms a reversible complex with the double helix form of DNA, as can be seen from the change in its ultraviolet spectrum in the presence of DNA as well as from the changes in the density, viscosity, heat denaturation curves, and ionic stability of DNA in solution in the presence of actinomycin. There are two models of its mode of combination with DNA. One is intercalation between the bases, as for daunomycin, and the other is insertion into the minor groove of the double helix of DNA. In any case, the actinomycin–DNA complex does not permit RNA polymerase to travel along the DNA template and thus inhibits synthesis of RNA. The increased stability of the double helix of DNA complexed to actinomycin is responsible for the inhibition of DNA replication.

a3. Mitomycins   These are antibiotics that bind covalently to DNA to form bridges between the two strands and impede their separation. Because of the nature of the bond the effects are irreversible, and therefore these agents are bactericidal. It is obvious that they are very toxic and cannot be used as antimicrobial agents in medicine.

b. Substances That Modify DNA

b1. Bleomycins   These form complexes with DNA, but unlike the other inhibitors, their effects seem to be due to the breaks they induce in the nucleotide chains. These antibiotics are used only against some types of tumors.

2. Inhibitors of RNA Polymerase

RNA polymerase inhibitors are much rarer than inhibitors of the template function. They have the following common characteristics:

1. Since prokaryotic and eukaryotic RNA polymerases are quite different, these inhibitors are usually selective. They inhibit only bacterial cells or animal cells, but not both.
2. They specifically inhibit RNA synthesis in growing bacteria, without any direct effect on DNA synthesis. The block of RNA synthesis causes protein synthesis to stop after a few minutes because of the absence of messenger RNA.
3. The temporary suspension of RNA synthesis is not in itself lethal to the cells. Therefore, inhibitors of RNA polymerase are bacteriostatic unless the bond formed with the enzyme is for practical purposes irreversible, in which case they are bactericidal.

a. Antibiotics That Inhibit RNA Polymerase

a1. Ansamycins   The ansamycins are antibiotics whose structure includes an aromatic group surmounted by a bridge consisting of an aliphatic chain and called the "ansa." The ansamycins that contain naphthalene aromatic

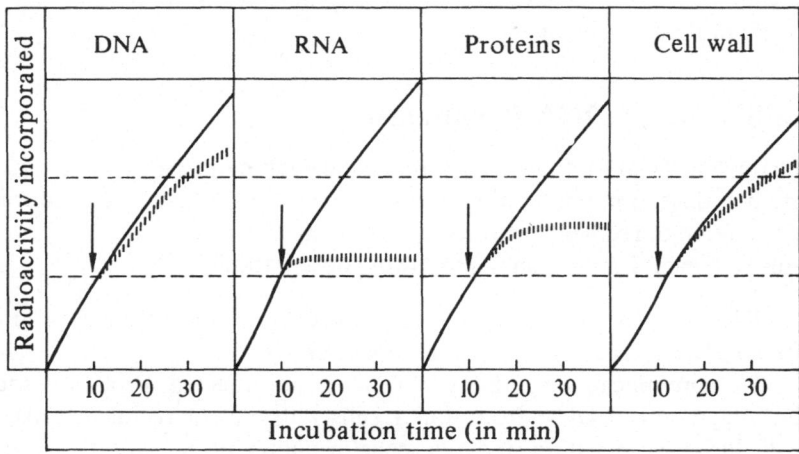

Figure 3.12. Effect of rifampin on incorporation of radioactive precursors into DNA, RNA, proteins, and cell walls of growing bacteria. The *solid line* indicates incorporation into a control culture, the *dashed line* incorporation into a culture to which antibiotic was added at the time indicated by the *arrows*.

groups (rifamycins, streptovaricins, tolipomycins, halomycins) are inhibitors of bacterial RNA polymerase.

Rifamycins are the best known of these, especially rifampin, a semisynthetic derivative with wide clinical use.

The action of the rifamycins results from the formation of a practically irreversible complex with RNA polymerase, as has been demonstrated by (1) the effect of these antibiotics on macromolecular synthesis in intact cells, shown in Fig. 3.12; (2) in vitro inhibition of bacterial RNA polymerase, but not DNA polymerase and not eukaryotic RNA polymerase; (3) the formation of a RNA polymerase–rifamycin complex demonstrable by chromatographic methods; (4) the inability of this complex to form with RNA polymerase extracted from rifampin-resistant bacteria.

Additional studies have shown that the antibiotic is bound to the β-subunit of the enzyme. Rifampin apparently inhibits the initiation process of RNA synthesis, not its elongation, as shown by the fact that the lengths of the few chains of RNA formed in the presence of the antibiotic are equal to the lengths of chains formed in its absence. More precisely, rifampin seems to block RNA synthesis immediately after the formation of the first phosphodiester bond.

a2. Streptolidigin   This is an antibiotic with poor activity, and therefore not used clinically, that inhibits bacterial RNA polymerase, as do the ansamycins. But its mechanism of action is different from that of the rifamycins because it acts on the elongation and on the initiation step of synthesis of the RNA chain.

3. Inhibitors of DNA Polymerase

Nalidixic acid (Nal) inhibits DNA replication taking place at the growing point, but does not inhibit the repair-type synthesis that occurs during recombination or recovery from radiation damage.

This statement is based on the following observations:

1. DNA synthesis is the first macromolecular synthesis to stop after addition of Nal to a culture of sensitive bacteria.
2. The nonreplicative synthesis of DNA is not affected. Processes such as the integration of DNA from donor cells into that of recipient cells of *E. coli* during conjugation and the transformation of *B. subtilis* are not inhibited by Nal, the DNA synthesis that follows ultraviolet treatment of *E. coli* is not affected by Nal even at high concentrations.
3. Nal does not interfere with nucleotide biosynthesis, since cell-free DNA-synthesizing systems using exogenously added triphosphate nucleotides are still sensitive to Nal.
4. Gyrase has been recently shown to be the target enzyme for Nal.

In the presence of Nal, sensitive cells become elongated, extensive DNA degradation occurs, and the cells die. The bactericidal effect of Nal is exerted only on bacteria actively synthesizing DNA and protein. The synthesis of these two macromolecules is essential for DNA degradation to occur. When cells treated with Nal are washed, DNA synthesis is restored, followed by restoration of growth. Therefore, Nal is not tightly bound to its target.

Conflicting evidence has been reported on the mode of action of two related antibiotics, novobiocin and coumermycin. It seems clear now that the primary effect on growing cells is inhibition of DNA synthesis, and it has been recently shown that gyrase is the target enzyme of these antibiotics. Nalidixic acid and novobiocin inhibit two different subunits of gyrase and thus they do not show cross-resistance.

4. Nonspecific Inhibitors of the Polymerizing Enzymes

Recently some semisynthetic rifamycins have been obtained that appear to have lost the specificity of action characteristic of the originals and to inhibit with approximately equal effectiveness both DNA polymerase and RNA polymerase obtained from completely different organisms, such as viruses, bacteria, and animal cells.

Their mechanisms of action are not yet clear. We know, however, that they seem to inhibit preferentially initiation of synthesis of the nucleotide chains, not elongation.

V. Inhibitors of Protein Synthesis

A. Phases of Protein Synthesis

Genetic information is translated by a process of protein synthesis in which the amino acids are polymerized according to an order determined by the sequence of the nucleotide triplets on the messenger RNA. These nucleotide sequences are, in turn, determined by the sequence of deoxynucleotides in specific tracts of the DNA.

Protein synthesis can be divided into two principal processes: (1) activation and identification of the amino acids by means of attachment of these to transfer RNA (tRNA) molecules and (2) polymerization of the amino acids, which takes place on the ribosomes and is also known as the ribosomal cycle.

1. Activation and Identification of the Amino Acids

There is a specific enzyme for each amino acid, an aminoacyl-tRNA synthetase, that catalyzes the formation of an ester linkage between the carboxyl of the amino acid and a hydroxyl on the last nucleotide of its specific tRNA.

Energy for formation of the bond is provided by ATP. The amino acid is activated, since the energy in the ester bond is sufficient for formation of the peptide bond, and identified by being attached to the proper tRNA. This process is also called charging of the tRNA and the resulting molecule is called aminoacyl-tRNA. In bacteria a specific aminoacyl-tRNA is formylated on the amino group and becomes the tRNA that initiates the polymerization process (tRNA initiator).

2. Ribosomal Cycle

a. Initiation Phase

The synthesis initiates with the formation of a complex between messenger RNA, the tRNA initiator, and the 30S subunit of the ribosome. The 50S subunit then joins this complex to complete the ribosome. The tRNA initiator binds to the first AUG triplet of the messenger RNA, in a position on the ribosome called site A. A movement of the ribosome (transloca-

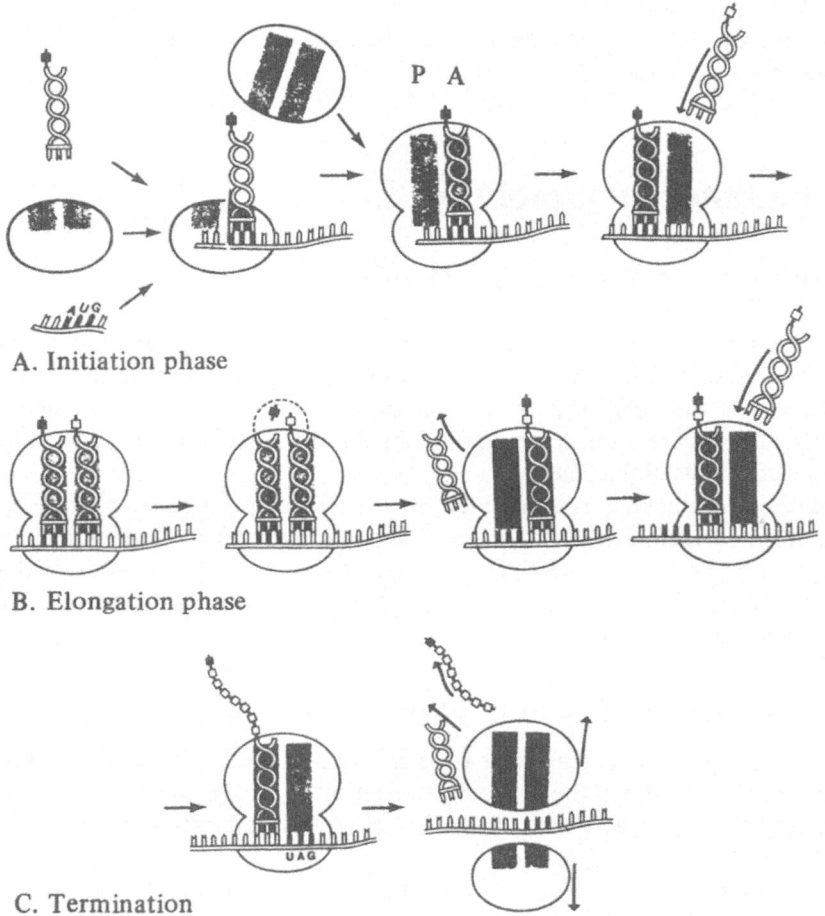

A. Initiation phase

B. Elongation phase

C. Termination

Figure 3.13. Ribosomal protein synthesis cycle. (Modified from M. Vergnano and D. Sassella (eds.), *Rassegna Medica*. Milan, Italy: Gruppo Lepetit Spa., 1973.)

tion) (for which energy provided by GTP is needed) causes the tRNA to move to another position (called site **P)** and site A is again free (Fig. 3.13A).

b. Elongation Phase

A new aminoacyl-tRNA attaches to site A, using energy provided by GTP and catalyzed by proteins called elongation factors (Fig. 3.13B). At this point, there are two aminoacyl-tRNAs attached to the ribosome, on specific messenger triplets and in sites A and P. The ester bond between the amino acid and tRNA at site P is broken and the free carboxyl forms a peptide bond with the amino acid on the tRNA at site A. This reaction is catalyzed by the enzyme peptidyl transferase, which is one of the proteins of the 50S subunit. The discharged tRNA leaves the ribosome and site P is freed. There is again a translocation of the ribosome, moving the initial chain (peptidyl-tRNA) from site A to site P, and a new aminoacyl-tRNA can attach to site A. This process is repeated as the peptide chain elongates.

c. Termination

The process continues until the ribosome arrives at a specific triplet on the messenger that is the signal for termination (Fig. 3.13C). At this point, the complex of peptidyl-tRNA, ribosome, and mRNA dissociates and the components start a new cycle. In the process of termination it appears that protein termination factors also play a role.

B. General Properties of the Inhibitors of Protein Synthesis

Antibiotics inhibit protein synthesis by different mechanisms and at different levels. Obviously, their overall effects on the cell are not identical. However, some generalizations can be made.

1. Temporary arrest of protein synthesis is not per se lethal to the cell. In fact, it always occurs when, for example, a bacterial culture lacks essential nutrients. Therefore, inhibitors of protein synthesis are bacteriostatic if they do not form irreversible bonds with some essential component of the synthetic system. If they do, they are bactericidal.

2. The effects of stopping protein synthesis on the synthesis of other macromolecules are complex (Fig. 3.14): (a) replication of DNA already initiated continues to completion of the chromosome, but new replication cannot be initiated; (b) RNA synthesis continues for a time, and may even in some cases be stimulated if the antibiotic acts at the ribosomal level; if, however, an antibiotic inhibits the charging of tRNA, in most bacteria there will be in addition to the blocking of protein synthesis a blocking of RNA synthesis as a secondary effect; (c) the rate synthesis of the cell wall declines and eventually ceases entirely.

3. The process of protein synthesis is more or less the same in all organisms. However, some components do differ in different taxonomic groups. The principal differences between prokaryotic and eukaryotic cells are listed in Table 3.2.

Protein-synthesis inhibitors have selective effects when they inhibit structures or components that differ in the various groups of organisms and nonselective effects when they inhibit components common to all cells.

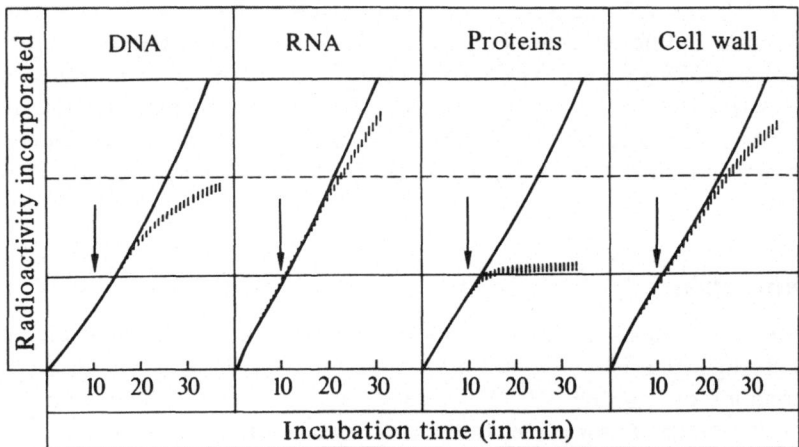

Figure 3.14. Effect of chloramphenicol on incorporation of radioactive precursors into DNA, RNA, proteins, and cell walls of growing bacteria. The *solid line* indicates incorporation into a control culture, the *dashed line* incorporation into a culture to which antibiotic was added at the time indicated by the *arrows*.

Therefore, there are selective inhibitors of protein synthesis in the prokaryotes, selective inhibitors of protein synthesis in eukaryotes and nonselective inhibitors. The antibiotics in the last two classes are obviously toxic, except for some special cases, such as the tetracyclines, described later. One must also remember that in the eukaryotic cells structures such as the mitochondria have special protein-synthesizing systems that resemble those of bacteria. Antibiotics that act selectively on the prokaryotic

Table 3.2. Differences between Prokaryotes and Eukaryotes in Protein Synthesis

	Eukaryotes	Prokaryotes
Ribosomal size	80S	70S
Ribosomal composition		
Protein	60%	40%
RNA	40%	60%
Ribosomal RNA	5S, 29S, 18S[a]	5S, 23S, 16S
	5S, 25S, 16S[b]	
Initiator aminoacyl-tRNA	met-tRNA$_I$	F-met-tRNA$_I$
Initiator factors		
Elongation factors	different in the two groups	

[a] Animals.
[b] Plants.
S, Svedberg unit, a measurement of the sedimentation coefficient.

system may, therefore, inhibit protein synthesis in these organelles and thereby cause side effects in eukaryotic cells.

With regard to the ribosomal sites, usually antibiotics that interfere at site P are selective, whereas those that interfere at site A inhibit nonspecifically both eukaryotes and prokaryotes.

C. Antibiotics that Inhibit Protein Synthesis

The antibiotics that inhibit protein synthesis are a large and diverse group of substances, some of which have important clinical applications. They may be conveniently divided into four subgroups, according to their sites of action:

1. inhibitors of amino acid activation and transfer reactions;
2. inhibitors of functions of the smaller (30S) ribosomal subunit;
3. inhibitors of the functions of the larger (50S) ribosomal subunit;
4. inhibitors of extraribosomal factors.

No clinically important antibiotic is a member of the first subgroup, a representative of which is borrelidin, an inhibitor of the transfer of activated threonine to its specific tRNA.

1. Inhibitors of Functions of the 30S Subunit

The 30S ribosomal subunit has two major functions: (1) It provides a site of attachment for mRNA; (2) it provides a site of attachment for the initiator aminoacyl-tRNA (N-formyl-methionyl-tRNA) and for the subsequent amino-acyl-tRNA (acceptor or A site). Therefore, one would predict that drugs binding to the 30S ribosomal subunit could inhibit initiation of protein synthesis and result in a block of the acceptor site for elongation. The most important antibiotics acting on 30S ribosomal subunits are the aminoglycosides and the tetracyclines.

a. Streptomycin and Other Aminoglycoside Antibiotics

The statement that streptomycin interferes with 30S subunit functions is based on the following experimental results: (1) Radioactive streptomycin binds to purified 30S subunit but not to 50S; (2) 30S subunit extracted from streptomycin-resistant mutants do not bind radioactive streptomycin; (3) none of the functions associated with the 50S subunit is inhibited by streptomycin. The analysis of ribosomal subunits extracted from resistant mutants has shown that streptomycin binds to a protein called P10. The binding of the drug with this protein causes a distortion of the A site of the 30S subunit, with consequent inhibition of correct positioning of the aminoacyl-tRNA.

Other aminoglycoside antibiotics also bind to the 30S subunit, at a site close to but not identical with that of streptomycin.

Streptomycin produces "misreading" in cell-free synthesis; i.e., it induces the incorporation of wrong amino acids into the growing polypeptide chain. It has been claimed for some time that the potent bactericidal effect of streptomycin is due to the synthesis of nonfunctional proteins. However, this is unlikely to be true, since addition of streptomycin to a growing culture causes a complete cessation of protein synthesis within a few minutes, a time insufficient for the synthesis of a substantial amount of nonfunctional proteins. The bactericidal effect of the drug is, rather, a consequence of the high affinity of streptomycin for its ribosomal site, which results in irreversible binding and irreversible ribosomal inactivation.

b. Tetracyclines

Evidence that the 30S is the site of action of the tetracylines includes the following experimental observation: (1) More radioactive tetracycline binds to 30S than to 50S subunits; (2) tetracycline inhibits the binding of radioactive F-met-tRNA$_F$ to 30S in the presence of the initiator codon AUG; (3) tetracycline inhibits the binding of aminoacyl-tRNA to 30S; (4) tetracycline does not interfere with the specific functions of the 50S ribosomal subunit. As "tetracycline-resistant" ribosomes have never been isolated, the precise site of its action has not been identified. Tetracycline inhibits the function of the smaller ribosomal subunits (40S) of the eukaryotic organisms, when tested in a cell-free system. However, (as already mentioned) it is inactive against intact eukaryotic cells, so the basis of its selectivity resides in the differential permeability. This antibiotic concentrates in the prokaryotic cell to inhibitory levels but is excluded by the membrane of eukaryotes.

2. Inhibitors of 50S Ribosomal Subunit Functions

The 50S ribosomal subunit has two major functions: (1) It provides a site for the attachment of the peptidyl-tRNA (donor site or "P" site), and (2) it participates in the formation of the peptide bond. Once the peptide bond is formed, the peptidyl-tRNA moves along the ribosome to make the acceptor site available to the next amino acid-tRNA (translocation).

The most important antibiotics acting on the 50S ribosomal subunits are discussed below.

a. Puromycin

Puromycin can be viewed as a structural analog of the 3' terminal of an aminoacyl-tRNA. As such, it binds to the acceptor site (A) of the smaller ribosomal subunit of both prokaryotes (30S) and eukaryotes (40S). The amino group of this antibiotic reacts with the free carboxyl end of the peptidyl-tRNA to form peptidyl-puromycin, which falls off the ribosome. Protein synthesis is thus blocked at the elongation step, and small incomplete peptide chains with a molecule of antibiotic covalently bound to the carboxyl end are formed. The lack of prokaryote versus eukaryote selectivity precludes the clinical use of puromycin.

The "puromycin reaction," i.e., the formation of peptidylpuromycin catalyzed by 50S ribosomal subunits, has been extensively used in studies of the mechanism of action of protein synthesis inhibitors. Because peptide bond formation is a prerequisite for the puromycin reaction, any inhibitor of the former causes inhibition of the latter.

The enzyme responsible for the peptide bond formation (often referred to as peptidyl transferase) is an integral part of the 50S (and 60S) ribosomal subunit. The puromycin reaction also occurs when, instead of peptidyl-tRNA, a peptidyl-oligonucleotide (such as peptidyl-CAACCA) is used. This peptidyl-oligonucleotide is equivalent to a 3' fragment of peptidyl-tRNA. (For this reason the reaction between puromycin and peptidyl-CAACCA is called the fragment reaction.)

The advantage of the fragment reaction is that it is limited to that part of the 50S subunit in the immediate vicinity of the catalytic center for peptide-bond formation, since both substrates are of small size and lack the sites for interaction with the ribosomes that are present in the tRNA. Therefore, inhibition of the fragment reaction may be taken as evidence of specific action on the peptide-bond formation center (peptidyl transferase).

b. Macrolides

The antibiotics of this group (of which erythromycin is the most extensively studied) inhibit protein synthesis by interfering with some phases of translocation or by distorting the conformation of the peptidyl transferase region. This conclusion is based on the observation that erythromycin inhibits chloramphenicol binding to the 50S subunit and that it binds to 50S ribosomal subunits extracted from sensitive bacteria more tightly than to those extracted from resistant ones. Other macrolides have been shown to bind to a ribosomal site partially coincident with that of erythromycin.

c. Chloramphenicol (CAF)

A description of the mechanism of action of chloramphenicol is given to illustrate the logical sequence of experiments needed to elucidate the mode of action of a protein-synthesis inhibitor.

CAF specifically inhibits protein synthesis in intact prokaryotic cells. The addition of CAF to a growing culture of sensitive bacteria causes a rapid cessation of protein synthesis, while the syntheses of DNA, RNA, and peptidoglycan continue almost unaltered for some time. This indicates that protein synthesis is the primary target of CAF action.

CAF interferes with ribosomal functions, a conclusion based on the following experimental results: (1) CAF inhibits amino acid polymerization in a cell-free system composed of aminoacyl-tRNA, ribosomes, mRNA, soluble factors, Mg^{2+} and GTP. This result excludes the possibility that CAF acts on amino acid activation and transfer reactions, and suggests that it acts on either a ribosomal or a soluble factor function. (2) Radioactive CAF binds to the ribosome, suggesting that this particle is its specific target.

CAF binds to 50S but not 30S ribosomal subunits. It is possible to separate the two ribosomal subunits. Preparations of purified 50S subunits bind radioactive CAF, whereas 30S do not.

CAF inhibits peptide bond formation. It is a potent inhibitor of the puromycin reaction carried out with leu-acetyl-CAACCA as substrate (fragment reaction). Therefore, it must interfere with the peptidyl transferase.

CAF binds onto the acceptor site of the 50S subunits. The drug inhibits the binding of aminoacyl-CAACCA but not that of peptidyl-CAACCA, suggesting that its binding site lies in the vicinity of the acceptor (aminoacyl) site.

d. Lincomycin and Clindamycin

The mechanisms of action of these antibiotic are very similar to that of erythromycin, and they demonstrate partial cross-resistance with it. Like chloramphenicol, lincomycin acts on a site that is at least partly coincident with that of erythromycin. Lincomycin also blocks elongation of the peptide chain.

3. Inhibitors of Extraribosomal Factors

These antibiotics interfere with the function of the soluble factors involved in protein synthesis.

a. Fusidic Acid

This steroidal antibiotic interferes with the function of the elongation factor (EF-G). In its presence, the ternary complex EF-G—GDP—ribosome is stabilized, ribosomal translocation is blocked, the EF-G factor is "frozen" in the ternary complex, and therefore is no longer available for another translocation cycle (Fig. 3.13). This conclusion is based on the following experimental observations: (1) Radioactive fusidic acid binds to purified EF-G factor (in a 1:1 molar ratio); (2) it inhibits the ribosomes-dependent GTPase activity of factor EF-G; (3) EF-G extracted from *E. coli* mutants resistant to fusidic acid has GTPase activity that is resistant to the antibiotic. Fusidic acid exerts the same inhibitory action on the eukaryotic elongation factor EF-2. Its in vivo selectivity toward prokaryotes (essentially Gram-positive bacteria) is based on selective permeability.

b. Kirromycins

Recent studies have shown that the antibiotics of this family inhibit the elongation process by binding to elongation factor Tu (EF-Tu) whose functions include the correct positioning of tRNAs on ribosomes.

The binding induces an allosteric alteration of the enzyme structure affecting all the reactions involved in the formation of the complex aminoacyl-tRNA–EF-Tu–GTP.

VI. Inhibitors of Cell Membrane Functions

All cells, both prokaryotic and eukaryotic, are surrounded by a cell membrane (or cytoplasmic membrane) that separates them from the external environment and that controls the passage of substances between the exterior and the interior. Although the supermolecular structure of the cell membrane is not completely known, it is known that it consists of a double lipid layer in which molecules of protein are intercalated, as in a mosaic (Fig. 3.15).

The membranes have very similar constituents throughout the phylogenetic ladder from bacteria to mammalian cells, but with one important difference. There are no sterols in bacterial cell membranes; zymosterol and

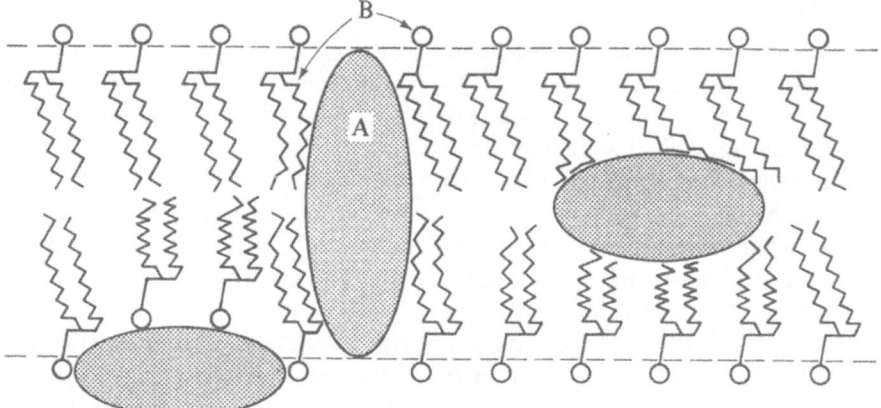

Figure 3.15. Schematic representation of the structure of the membrane by the classic "mosaic" model. (**A**) Protein; (**B**) lipids.

ergosterol are present in the cell membranes of fungi and plants, and cholesterol in those of mammals.

Antibiotics that act on the cell membrane can be divided into two groups:

1. substances that disorganize the supermolecular structure of the membrane, thus causing loss of cellular substance to the outside;
2. substances that act as carriers for specific ions (for this reason called ionophores) and cause an abnormal accumulation of ions inside the cell.

In each case, the antibiotics that act on the membrane cause a loss of the osmotic properties of the cell.

A. General Properties

Although antibiotics that act on the membrane do so via different mechanisms, they have the following properties in common:

1. The majority are poorly selective, acting against both bacterial cells and those of higher organisms. This is easy to understand when one remembers the considerable chemical and structural similarities of the cell membranes of different organisms.

2. As a consequence of this lack of specificity, they are usually too toxic to be given systemically and are used almost exclusively by topical application.

B. Antibiotics Active Against the Membrane

The antibiotics that disorganize the structure of the cell membrane are essentially of two types:

1. Lipopeptide substances consisting of a lipophilic segment and a hydrophilic segment that can insert themselves between the lipid and the protein of the membrane structures, irreversibly altering the structures (polymyxin, colistin, circulin).
2. Substances that form complexes with the sterols of the membrane and thus change the structure. Since the prokaryotes do not contain sterols in their membranes, they are not susceptible to these antibiotics, which act against yeasts, fungi, and animal cells. The antibiotics of this group have polyene structures, consisting of an aliphatic chain closed into a ring and containing a lipophilic zone with a series of conjugated double bonds and a hydrophilic zone with a series of hydroxyl groups.

   Some polyenes interact with greater affinity with ergosterol (the sterol in fungal membranes) than with cholesterol (the sterol in animal cell membranes) and therefore have a certain selective toxicity for fungi and can be used as antifungal agents. The two most important of these antibiotics are amphotericin B and nystatin.

Ionophore antibiotics are a large and diverse group of molecules with the capacity to form lipid-soluble complexes with cations, which can then cross the lipophilic cell membrane. Therefore, they alter the permeability of specific cations. The different ionophores differ in (1) the molecular mechanisms by which they alter transmembrane cation permeability; (2) specificity toward different cations. In terms of the mechanism of action these antibiotics can be grouped in the following classes.

1. Stationary Ion-Conducting Channels

These molecules insert themselves into the structure of the cell membrane, forming channels whose walls are lipophilic on the outside and hydrophilic on the inside, thus permitting cations to leak from the cell. Members of this group are the linear peptides gramicidins A, B, C, and the cyclic peptide alamethacin. They lack prokaryote–eukaryote specificity and therefore cannot be used as systemic antibacterial drugs. Gramicidins, which have low MICs against certain Gram-positive bacteria, are used topically in some countries.

2. Mobile Ionophores

Mobile ionophores are molecules that can complex with ions and are able to move through the cellular membrane, thus inducing an abnormal transcellular distribution of specific cations. Chemically they belong to three major classes.

a. Cyclodepsipeptides

Cyclodepsipeptides include such antibiotics as valinomycin and the enniantins. Valinomycin (Fig. 3.16) in solution assumes a ring-like conformation, with the ether carbonyls forming a rigid inner sphere that can easily accommodate K^+ but not the smaller Na^+ or Li^+. This is the basis of its very high selectivity toward alkali cations: It discriminates between K^+ and Na^+ in a ratio of $10^4:1$. The enniantins are smaller than valinomycin and therefore can accommodate the smaller alkali cations. The K^+ versus Na^+ selectivity is only $10:1$.

b. Macrotetrolides

These are molecules made up of four tetrahydrofuranyl hydroxyacids linked together through a lactone linkage. Like the cyclodepsipeptide ionophores, they can accommodate K^+ and function as mobile K^+ carriers.

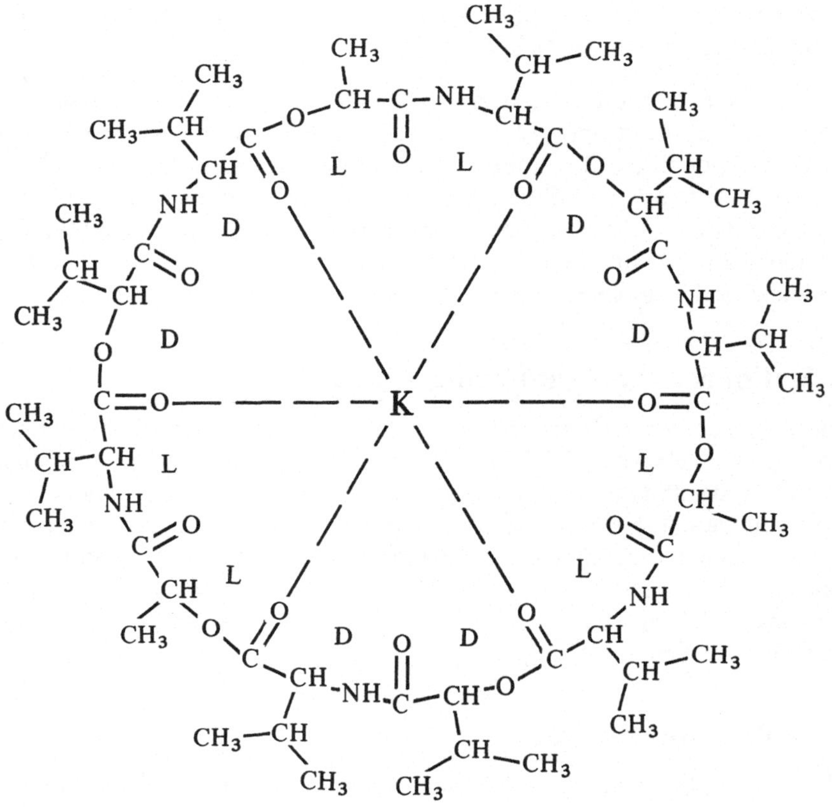

Figure 3.16. Schematic representation of the interaction between valinomycin and K^+.

c. Sideromycins

Sideromycins are antibiotics that interfere with the uptake of iron by the microbial cells and thus cause cellular iron deficiency that results in suppression of growth.

Many bacterial pathogens acquire the iron they need for growth through a complex mechanism that involves several steps:

1. An iron transport cofactor (siderophore) is synthesized.
2. It is then excreted outside the cell into the biological fluid of the host.
3. Since the affinity constant for iron is of the same order of magnitude as that of the transferin of serum, some iron is extracted from this serum protein.
4. The iron–siderophore complex (usually in the ferric form) is reassimilated into the cell through a transport mechanism that involves specific carriers localized in the cell membrane.

5. The iron is liberated as needed from the chelate by reduction of Fe^{3+} or by enzymatic hydrolysis of the ligand.

The sideromycins are believed to compete with the siderophores for the transport carriers and thus to inhibit the assimilation of iron by the cells. They can be considered "antagonists," i.e., competitive inhibitors, of the siderophores. They also compete in the desferri-form, i.e., in the form that does not contain iron. The frequency of resistant mutants in a given population is very high (10^{-3}–10^{-5}), which practically precludes the use of sideromycins as therapeutic agents. The nature of the biochemical alteration in the mutants is not known.

Siderophores are produced and presumably utilized as iron-transport cofactors by a variety of free-living microorganisms, including streptomycetes, mycobacteria, and fungi. The chemical structures of several such compounds (earlier called syderamines) have been elucidated.

VII. The Antimetabolites

The term antimetabolites refers to a group of natural and synthetic substances, with very heterogeneous chemical structures and mechanisms of action, but having in common the fact that their inhibitory effects can be antagonized by one or more metabolites. Generally, but not always, their chemical structures are analogous to those of the antagonistic metabolites.

A. Mechanism of Action

The antimetabolites can be divided into two large groups on the basis of their mechanisms of action:

1. those that are incorporated into "informational" polymers (DNA, RNA, proteins) in place of natural monomers and change the information content;
2. those that inhibit the formation of essential metabolites (Fig. 3.17).

In both cases, they have the following properties:

1. They are often structurally similar to natural metabolites, of which they are said to be analogs.
2. They interact with enzymes that normally recognize the natural metabolite and are thus made unavailable to the natural metabolite.
3. The degree of inhibition is a function of the ratio of the concentration of the analog to that of the natural metabolite; this type of inhibition is called competitive.

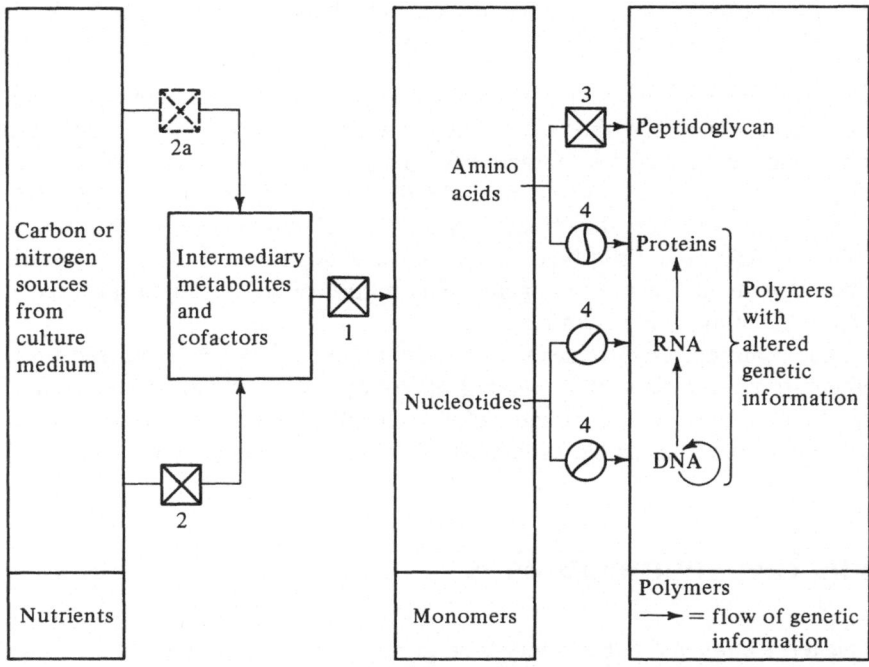

Figure 3.17. Metabolic sites of actions of antimetabolites: (*1*) block of monomer synthesis with consequent block of polymer synthesis and hence of bacterial growth; (*2*) block of synthesis of cofactors essential for monomer
synthesis; it is also possible to imagine the existence of substances that block
the synthesis of one or more intermediary metabolites essential for the synthesis of monomers (*2a*); because these metabolites are produced by synthetic
pathways that are more or less the same through the evolutionary scale, such
substances would not be selective and would be toxic; (*3*) block of polymerization; (*4*) incorporation of analogs that make the polymers nonfunctional.

4. The effects of the antimetabolite can be overcome not only by the natural product to which they are analogous, but also by the final product
   of the inhibited metabolic pathway.

B. Biosynthetic Incorporation

The polymerization system for amino acids often cannot distinguish between the amino acid and its analog and incorporates either into the protein. In this way, abnormal proteins are synthesized that have little or no
function, and cell growth is inhibited. The best known amino acid analogs
acting in this way are fluorophenylalanine (a synthetic product) and seleno-
methionine (which occurs naturally in some plants). There are also analogs of the pyrimidine bases that are incorporated into DNA or RNA

instead of the natural compounds. The most important synthetic compound of this type is 5-bromouracil, which competes with thymine in DNA synthesis.

C. Inhibition of the Synthesis of a Metabolite or Essential Cofactor

Unlike the substances described in the preceding section, many antimetabolites are not incorporated into the macromolecules in place of the metabolites they resemble, but inhibit one or more enzymes in the synthetic pathways for metabolites. The final result is inhibition of the synthesis of the macromolecule of which that metabolite is an essential component.

The same result can be obtained if an antimetabolite interferes with the synthesis of a cofactor that is indispensable for the synthesis of one or more essential metabolites. Many antimetabolites with different effects on cellular functions (synthesis of intermediary metabolites, synthesis of amino acids, mechanisms of transport) have been described. The most important belong to the following groups:

1. inhibitors of nucleotide synthesis and hence of nucleic acid synthesis;
2. inhibitors of the synthesis of the pentapeptide of peptidoglycan, and hence of cell wall synthesis;
3. inhibitors of the synthesis or the function of folic acid, a cofactor that is indispensable for the synthesis of purine nucleotides and of proteins.

1. Inhibitors of the Synthesis of Purine and Pyrimidine Nucleotides

These include hadacidin and alanosine.

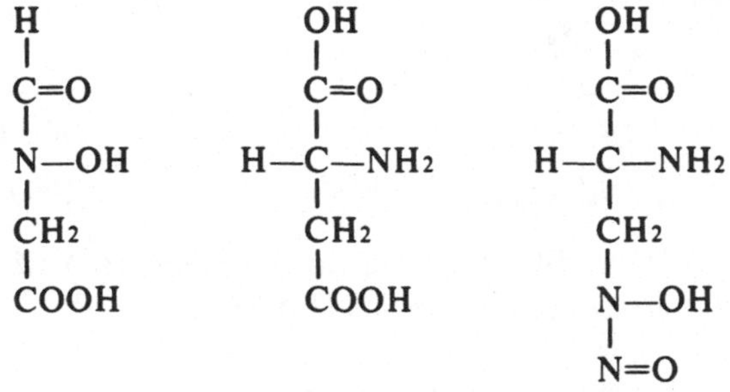

$$
\begin{array}{ccc}
\text{H} & \text{OH} & \text{OH} \\
| & | & | \\
\text{C=O} & \text{C=O} & \text{C=O} \\
| & | & | \\
\text{N—OH} & \text{H—C—NH}_2 & \text{H—C—NH}_2 \\
| & | & | \\
\text{CH}_2 & \text{CH}_2 & \text{CH}_2 \\
| & | & | \\
\text{COOH} & \text{COOH} & \text{N—OH} \\
& & | \\
& & \text{N=O} \\
\text{Hadacidin} & \text{L-Aspartic acid} & \text{Alanosine}
\end{array}
$$

a. Hadacidin

Hadacidin is an aspartic acid analog. Aspartic acid takes part in the synthesis of AMP by the following reactions: IMP + aspartic acid → adenylsuccinate. Hadacidin competes with aspartic acid for the site on the enzyme and inhibits the reaction in that way. This is a competitive type of inhibition, since it is a function of the ratio of the concentration of the analog (hadacidin) to that of the antagonist (aspartic acid). The same reaction is inhibited by alanosine, another antibiotic analog of aspartic acid.

A similar mechanism underlies the antibiotic effects of azaserine and diazoisonorleucine (DON), which are analogs of glutamine, an essential

$$
\begin{array}{ccc}
\text{COOH} & \text{COOH} & \text{COOH} \\
| & | & | \\
\text{H—C—NH}_2 & \text{H—C—NH}_2 & \text{H—C—NH}_2 \\
| & | & | \\
\text{CH}_2 & \text{CH}_2 & \text{CH}_2 \\
| & | & | \\
\text{O} & \text{CH}_2 & \text{CH}_2 \\
| & | & | \\
\text{C=O} & \text{C=O} & \text{C=O} \\
| & | & | \\
\text{CH} & \text{CH} & \text{NH}_2 \\
\| & \| & \\
\text{N}^+ & \text{N}^+ & \\
\| & \| & \\
\text{N}^- & \text{N}^- & \\
\text{Azaserine} & \text{DON} & \text{Glutamine}
\end{array}
$$

precursor in the synthesis of purine nucleotides. Both analogs competitively inhibit the synthesis of the nucleotides and thus of the nucleic acids (Figs. 3.18, 3.19).

b. Showdomycin

Showdomycin is a uridine analog that interferes with the phosphorylation

Showdomycin   Uridine

of UMP to UDP, an obligatory step in the conversion of UMP to CTP. This last is a precursor of nucleic acids (Fig. 3.20).

2. Inhibitors of Peptidoglycan Synthesis

a. Cycloserine

The mechanism of action of cycloserine was described in Section III.D.2. This antibiotic can be considered an antimetabolite that inhibits the formation of peptidoglycan.

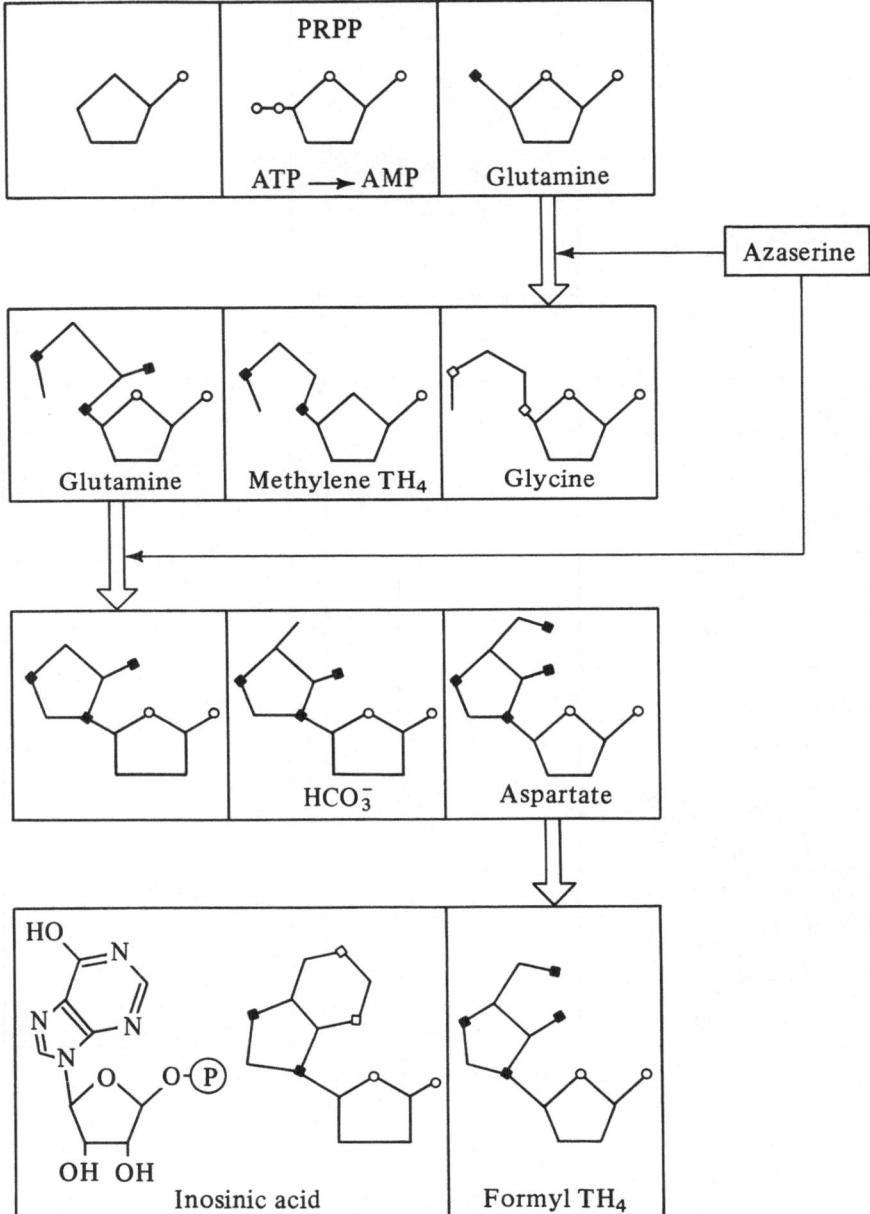

Figure 3.18. Site of action of purine nucleotide synthesis inhibitors. Initial steps.

Cycloserine is an "analog" of alanine and is antagonized by alanine. Its action can also be overcome by addition of D-alanine and D-analyl-D-alanine, which are final products in the metabolic pathway inhibited by cycloserine.

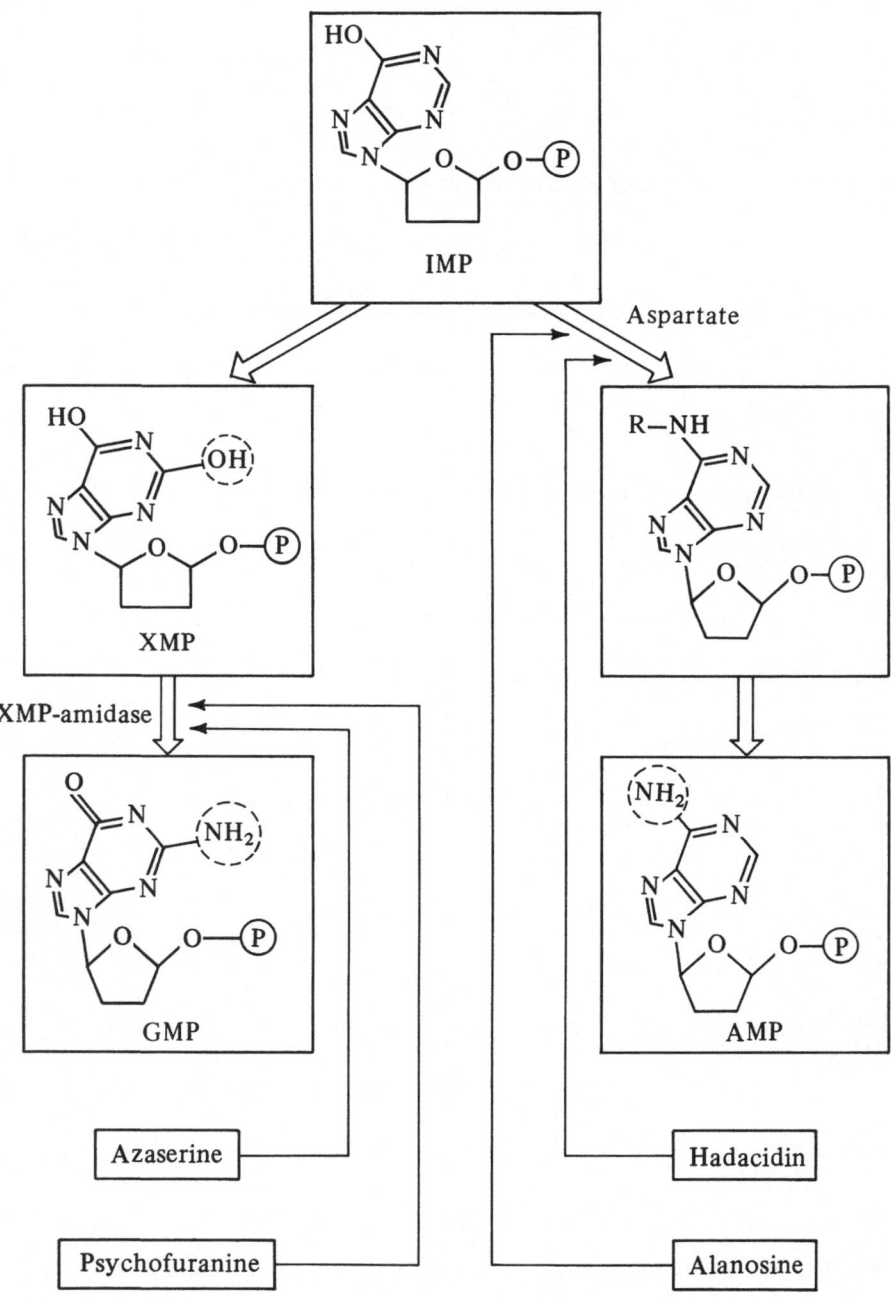

Figure 3.19. Site of action of purine nucleotide synthesis inhibitors. Final steps.

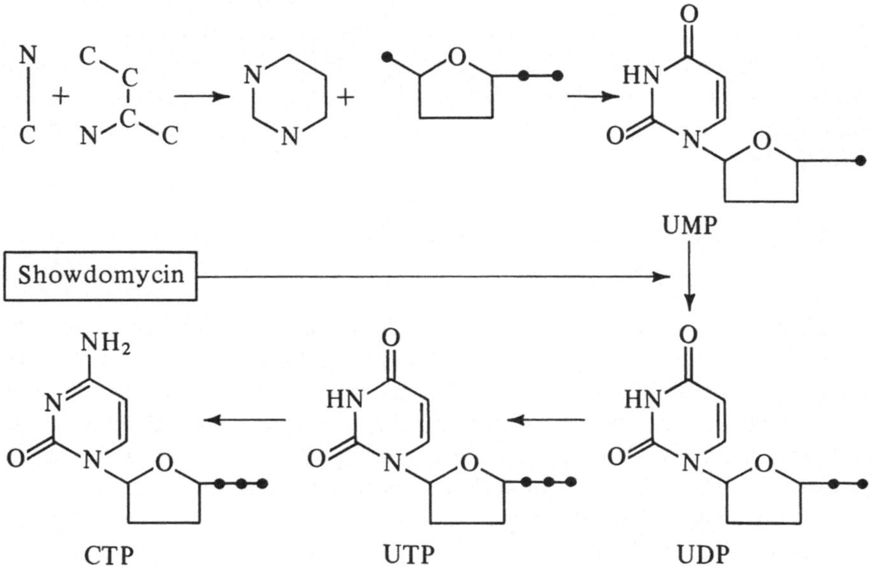

Figure 3.20. Site of action of showdomycin.

3. Antimetabolite Inhibitors of Folic Acid Synthesis

Although in this book we have limited ourselves to describing natural antibiotics, i.e., products obtained from microorganisms, it is worthwhile to mention at this point a group of synthetic antibacterial agents, the sulfonamides, because they illustrate one of the few cases of clinically useful

Sulfonamide   Para-aminobenzoic acid (PABA)

antimetabolites. It can be seen that the sulfonamides are structurally similar to PABA, which is a precursor of folic acid and, as a consequence, an essential cofactor for bacterial growth.

The sulfonamides competitively inhibit the incorporation of PABA into the folic acid molecule and are in part incorporated into molecules that resemble folic acid, forming folic acid analogs that are also inhibitors. Addition of folic acid to a bacterial culture inhibited by sulfonamides does not overcome the effects of the inhibitor, because the bacteria are impermeable to folic acid. If, instead, the final products of "folic acid metabolism," e.g., methionine or thymine, are added, they do overcome the sulfonamide effect. Folic acid is also an essential cofactor in cells of higher organisms that are not able to synthesize it and that must obtain it from outside. Therefore, these cells are permeable to folic acid.

Because these cells do not contain folic acid synthesizing pathways, the sulfonamides are inactive in them. The selectivity of sulfonamides is therefore attributable to: (1) the presence of a particular metabolic pathway sensitive to the sulfonamides, essential for bacteria but not present in mammalian cells; (2) the inability of the bacterium to transport folic acid inside the cell.

4. Inhibitors of Dihydrofolic Reductase (Trimethroprim)

Biochemical utilization of folic acid includes, among other reactions, its reduction to dihydrofolic acid and tetrahydrofolic acid, catalyzed by the enzyme dihydrofolate reductase. This enzyme is present throughout the evolutionary scale, from bacteria to protozoa to mammals, including man, but its structure differs in the different groups of organisms. Because of these structural differences, at certain concentrations some compounds will inhibit the bacterial enzyme but not that in protozoa or in mammalian cells. The reverse has also been found. This is a typical case of differential affinity. Trimethroprim (Tmp) is an inhibitor of the bacterial but not of the mammalian enzyme.

Tmp is a structural analog of the pteridine portion of dihydrofolic acid reductase (see Fig. 3.20). In the presence of Tmp, tetrahydrofolic acid is depleted and converted into dihydrofolic acid. As a consequence, the thymidilate synthetase reaction that converts deoxyuridilic acid to thymidilic acid stops. This reaction, in fact, uses tetrahydrofolic but not dihydrofolic cofactors.

Following thymidilate synthesis inhibition, DNA synthesis ceases and cell growth stops. Addition of thymine to the culture medium reverses the inhibitory action of trimethroprim.

The synthesis of other one-carbon metabolism products, in addition to that of thymidilate, is also inhibited by Tmp (glycine, methionine, purines) (see Fig. 3.21).

The mode of action of Tmp (Fig. 3.22) depends on the selective availability of one-carbon metabolism products. If none is present in the

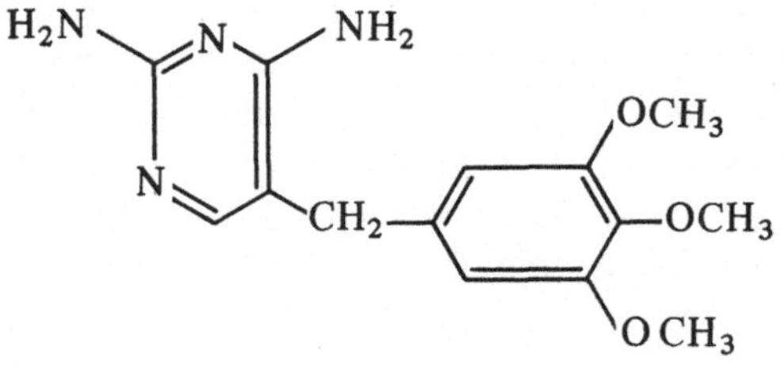

Figure 3.21. Chemical structure of trimethoprim.

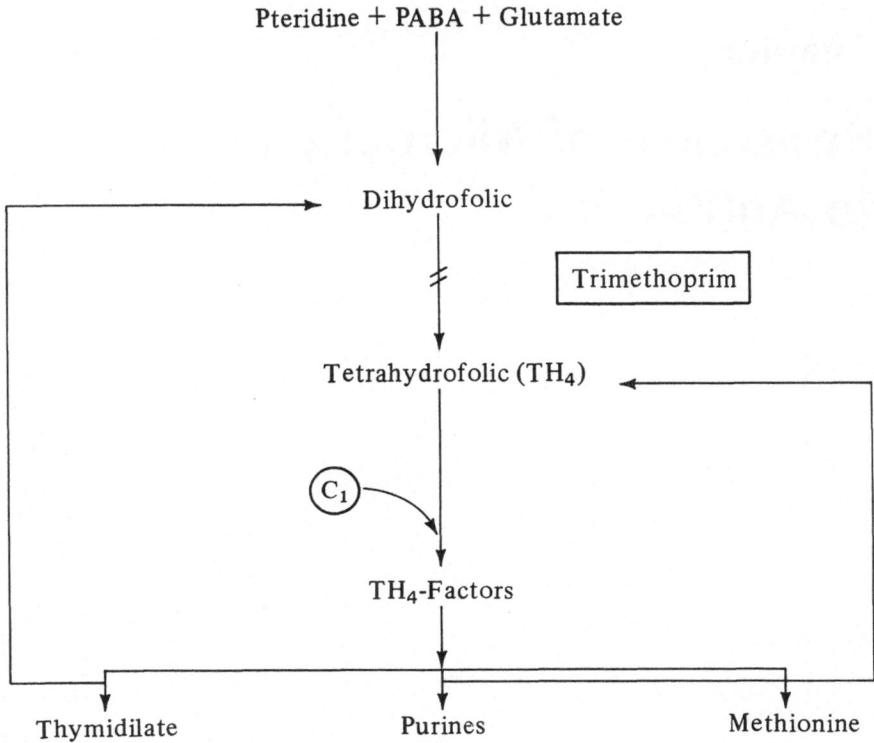

Figure 3.22. Mechanism of action of trimethoprim.

medium, the drug is bacteriostatic. If thymine is absent but glycine and methionine present, the drug is bactericidal (thymineless death). If both thymine and the amino acids are present, the drug has no inhibitory effect.

Chapter 4

Resistance of Microorganisms to Antibiotics

I. General Aspects

Many bacterial infections that until a few years ago could be cured with specific antibiotics are today resistant to treatment with the same antibiotics. This is a general phenomenon, as it has been observed in different parts of the world, at different times, and for numerous (almost all) antibiotics. Its clinical importance is obvious. Bacterial resistance to antibiotics has been very actively investigated, measured, and followed at the epidemiological level. In this chapter, however, we do not detail the epidemiology of resistance but rather describe the biological aspects. We attempt to answer the following questions:

1. What is the difference between resistant and sensitive (susceptible) bacteria?
2. What role does the antibiotic play in the appearance and diffusion of the resistance?
3. Is the resistance a genetic phenomenon; i.e., is the information for the resistance contained within the bacterial DNA?
4. If so, is it transferred from one bacterium to another? How?

Before answering these questions, it is necessary to provide some definitions and a brief description of the methods used to measure resistance.

A. Definitions

Resistance: A bacterial strain derived from a species that is susceptible to an antibiotic is said to be resistant when it is not inhibited by the minimal concentration of the antibiotic that inhibits the growth of that species (MIC). This strain may of course be inhibited by a higher concentration of the same antibiotic. Therefore, it is obvious that the concept of resistance applies to (1) a bacterial strain, (2) an antibiotic, and (3) a concentration. The physician considers a bacterial strain resistant to a given antibiotic if it can grow in the presence of a concentration equal to or greater than that which the antibiotic can reach in serum or tissue.

Multiple resistance: When a bacterium is resistant to two or more structurally unrelated antibiotics, one speaks of multiple resistance. Many antibiotics with similar chemical structures inhibit bacterial growth by the same mechanisms of action. Usually, a bacterial strain that is resistant to an antibiotic is also resistant to other antibiotics of the same family. This resistance is called either cross-resistance or co-resistance. Cross-resistance is of considerable practical importance. If an infection is due to a bacterium resistant to a certain antibiotic, one must be sure not to treat it with other antibiotics of the same class (with exceptions).

One-step and multistep resistance: It is possible to isolate resistant mutants from a population of bacteria susceptible to a given antibiotic by spreading a large number of cells on a solid culture medium containing a concentration of antibiotic several times greater than the MIC. The great majority of bacteria will be inhibited, but if a sufficiently large number of cells, for example 10^9 or more, are used for the inoculum, one may find resistant cells that will give rise to easily isolated colonies.

Frequently the bacteria in these colonies are resistant not only to the concentration of antibiotic in the medium on which they were grown but also to higher concentrations. This is called one-step resistance, and occurs in the case of certain antibiotics for which the resistance is an all-or-none phenomenon; i.e., the cells are either susceptible to low concentrations or are not inhibited even by high concentrations. A typical example is the resistance to streptomycin. In the case of other antibiotics, e.g., penicillin, it is not possible to isolate resistant cells by the method described above. Usually, bacterial populations contain mutants resistant to the MIC of these antibiotics, but sensitive to slightly higher concentrations. Therefore, it is extremely difficult to find a concentration of antibiotic in solid medium that will inhibit the growth of all the normal cells and not of the resistant ones. Resistants of this type are selected from liquid culture media by a method inappropriately called "training." According to this method, the bacterial strain is grown in concentrations of antibiotic slightly below the MIC to select the less sensitive cells. These then give a population that includes cells capable of growing in the presence of slightly higher concentrations of antibiotic. By repeated passages of this type and selecting from large populations in liquid culture, one can obtain strains that are notably more resistant. This type of gradual resistance, found for penicillin and several other antibiotics, has been given the name multistep.

Resistance and unsusceptibility: To clarify the terminology, it is well to reserve the word "resistant" for those bacterial strains that come from an originally susceptible species and have lost their susceptibility to a given antibiotic through mutation or some other change in genetic heritage of the type described later. Bacterial species that are "intrinsically" not inhibited by an antibiotic, for example, because they lack the structure on which the antibiotic acts, are called insensitive (or unsusceptible). There can also be sensitive (susceptible) bacteria that are not inhibited by an antibiotic because of environmental and culture conditions. For example, the MIC of streptomycin against *Staphylococcus aureus* is much lower at pH 8 than at pH 5. Under acidic conditions *Staphylococcus aureus* will therefore be practically insensitive to streptomycin, although it is not resistant in the sense of the strict definition above. In discussing these cases, some authors use the term phenotypic resistance because the phenomenon is not related to any change in the genetic inheritance of the bacterial population.

B. Testing for Susceptibility (or Resistance)

The resistance of a bacterium to an antibiotic can be determined in two ways.

1. *Disc method.* This is the method used to construct antibiograms as described in section 2.VII. It is used preferentially in the medical microbiology lab. The concentration of antibiotic in the disc is chosen so as to provide information about the susceptibility of the bacteria to concentrations of the antibiotic attainable in a patient's serum.
2. *Determination of the MIC.* This consists of determining the minimum inhibitory concentration of the antibiotic (see sections 2.III and 2.IV). It is a more quantitative method and is used preferentially in genetic and epidemiological studies.

II. Biochemical Bases of Resistance

An antibiotic inhibits bacterial growth if it (1) penetrates the bacterial cell, (2) interacts with a structure involved in an essential function, and (3) substantially inhibits this function. A bacterium becomes resistant to an antibiotic if at least one of these steps is no longer operative. This can occur as the result of one of the following principal biochemical mechanisms:

1. transformation of the antibiotic into an inactive form,
2. modification of the cell's target site for the antibiotic,
3. modification of the permeability of the bacterium to the antibiotic,
4. increased production of the structure inhibited by the antibiotic.

Table 4.1 lists the types of resistance and the mechanisms for the more common antibiotics.

A. Modification of the Antibiotic

The resistant bacterium synthesizes an enzyme capable of chemically transforming the antibiotic into an inacitve product. Among these enzymes, the most important are:

1. The peptidases, which hydrolyze peptide bonds, and in particular the β-lactamases that open the β-lactam ring of penicillin and cephalosporin. The β-lactamases produced by different resistant strains are not all identical. For example, the staphylococcal β-lactamase is not active against cephalosporins, whereas *E. coli* β-lactamases are active.
2. The acetyltransferases, which inactivate antibiotics by transferring an acetyl unit from an acetyl donor to a functional group on the antibiotic. An example is the conversion of chloramphenicol to acetyl- or diacetyl-chloramphenicol.
3. Phosphoryltransferase, which introduces a phosphate group on the antibiotic, thereby inactivating it. Streptomycin can be subject to this modification.
4. Adenyltransferase, which inactivates the antibiotic by transferring an adenyl group to it.

The inactivating enzymes can be constitutive or induced by the antibiotic. In the latter case, the cell that contains the gene for resistance possesses control mechanisms that enable it to refrain from synthesizing an enzyme, β-lactamase for an example, when there is no need for it, e.g., in the absence of the antibiotic. In the presence of the antibiotic, a mechanism called induction is set off and the cell synthesizes the inactivating enzyme. Note that the antibiotic does not cause resistance, but only induces the expression of resistance potentially present in the cell.

B. Modifications of the Cell's Target Site for the Antibiotic

Many antibiotics act by inactivating a protein molecule that can be called generically a receptor. The antibiotic binds to such a receptor, forming a more or less stable complex. One large class of resistant mutants is comprised of bacteria that, through a mutation, contain a receptor unable to

Table 4.1. Mechanism of Resistance to Some Representative Antibiotics

Antibiotic	Type of mechanism	Location of genetic determinant	Description of mechanism
β-lactam antibiotics (penicillins and cephalosporins)	Inactivation	Generally extra-chromosomal	β-lactamases that open the β-lactam ring. Some specific for penicillins or cephalosporins; others do not distinguish between substrates. Usually inducible in Gram-positives and constitutive in Gram-negatives.
	Intrinsic	Chromosomal	Methicillin resistance (S. aureus).
	Tolerance	Unknown	S. aureus still inhibited by penicillin (low MIC) but not killed (high MBC). Tolerant strains have high content of autolysin inhibitor.
Chloramphenicol	Inactivation	Extrachromosomal	Acetylation by an inducible enzyme.
	Modification of target	Chromosomal	Modification of ribosomal receptor.
Aminoglycoside antibiotics	Inactivation	Extrachromosomal	N-acetylation: Different bacteria have different enzymes that have different specificities for different antibiotics and attack different NH_2 groups. In all cases the acetyl donor is acetyl CoA. Phosphorylation: Different bacteria have different enzymes that have different specificities for different antibiotics and attack different OH groups. In all cases the donor is ATP. Adenylation: An enzyme transfers the adenyl moiety from ATP, liberating pyrophosphate, to the 2-OH group of $3-NH_2-3-$deoxyglucose of different antibiotics.

Table 4.1. (*continued*)

Antibiotic	Type of mechanism	Location of genetic determinant	Description of mechanism
Streptomycin	Modification of target	Chromosomal	Alteration of the ribosomal 30S subunit.
	Inactivation	Extrachromosomal	Similar to other aminoglycosides.
Kasugamycin	Modification of target	Chromosomal	Alteration of the 16S RNA of ribosomal 30S subunit.
Erythromycin	Modification of target	Chromosomal	Alteration of a protein of the ribosomal 50S subunit.
		Extrachromosomal	Methylation of the ribosomal RNA (*S. aureus*).
Rifamycins	Modification of target	Chromosomal	Alteration of the β-subunit of RNA polymerase.
Cycloserine	Cellular permeability	Chromosomal	The system that transports L-alanine and glycine (which is used to transport cycloserine) is modified.
Tetracyclines	Cellular permeability	Extrachromosomal	Decreased efficiency of transport. The resistance is partially induced.
Fusidic acid	Modification of target	Chromosomal	Alteration of elongation factor G of protein synthesis.
Phosphomycin	Cellular permeability		Alteration in the transport system for glycerophosphate or glucose-6-phosphate (which is used to transport phosphomycin).

bind the antibiotic, or less often, a receptor retains its function even after formation of the complex. Frequently, this difference consists of substitution of only one amino acid in the protein chain. Examples of this type are some mutants resistant to streptomycin with an alteration in the streptomycin-binding ribosomal protein P10. There are mutants resistant to erythromycin that have an alteration in protein P8 in the ribosomal 50S unit, which is the site of action of this antibiotic and of other macrolides, such as oleandomycin. Rifampin-resistant bacteria are another example. The DNA-dependent RNA polymerase (the target of rifamycin) isolated from these mutants is not capable of forming a complex with the antibiotic. In at least one case this has been shown to be due to an alteration in the β chain, one of the five proteins that make up the enzyme. Bacteria have also been isolated that are both resistant and dependent on the presence of the antibiotic for growth (e.g., streptomycin); they do not grow if the antibiotic is not present. This curious finding has been explained by postulating that the target protein in the resistant strains is inactive but becomes functional upon binding of the antibiotic.

C. Modification of the Permeability of the Cell to the Antibiotic

Mutations that render the bacterial cell impermeable to the antibiotic have been frequently hypothesized to be the cause of resistance. In some cases, however, it was shown that another mechanism was involved; and, in other cases, doubts remain because no conclusive evidence of this effect has been obtained. Like other organic molecules, antibiotics penetrate the cell membrane by one of two major mechanisms: (1) passive diffusion, or (2) specific active transport. When the physical properties of the antibiotic are compatible with passive diffusion, it is very difficult to conceive a mutation that makes the cell impermeable to the antibiotic. This would in fact imply a major change in the membrane structure that most probably would be lethal to the cell. When penetration is due to a specific transport mechanism, as in the case of tetracyclines or aminoglycosides, a specific carrier protein may be involved and an alteration of this is a likely mechanism of resistance.

D. Increased Production of the Enzyme Inhibited by the Antibiotic

Mutants of this type are quite frequent among organisms resistant to antimetabolites, such as 5-methyltryptophan or trimethoprim. In the first case more tryptophan is produced; in the second, more folic acid reductase is produced. Mutants resistant to cycloserine produce an increased amount of alanine racemase or D-alanyl-D-alanine synthetase.

E. Other Possible Mechanisms

The above four mechanisms, especially the first two, are encountered most frequently. Other mechanisms, however, can be hypothesized; e.g., metabolites that would decrease the effect of the antibiotic; utilization of an alternative metabolic pathway that is not inhibited.

F. Comments

From the preceding discussion, it is possible to draw certain general conclusions:

1. There are many biochemical mechanisms for resistance to antibiotics. Some of these (such as the second and fourth types described) are related to the mechanism of action of the antibiotic. Others (the first type) are independent of the antibiotic's mechanism of action.

2. For any given antibiotic there may be strains that are resistant through different mechanisms. For example, the resistance to streptomycin is sometimes due to an inactivating enzyme and sometimes to a modification of the site of its action.

3. When two antibiotics act at the same site, cross-resistance is frequent. Bacteria resistant as a result of modification of the site of action will usually be resistant to both antibiotics.

When the resistance to one of the two antibiotics has another basis (e.g., an inactivating enzyme), the other antibiotic may remain active and show no cross-resistance.

III. Genetic Aspects

A. The Role of the Antibiotic in the Transformation of a Bacterial Population from Susceptible to Resistant

During the process of duplication of genetic material, errors of replication may occur that lead to a change in the genetic information in the microorganism. Such modifications are called mutations. Some mutations may render the microorganism resistant to an antibiotic to which the wild-type (the organism prior to mutation) was susceptible. Mutations toward resistance usually occur with a very low frequency, of the order of 10^{-7}–10^{-9}, and therefore the fraction of resistant cells in a bacterial population is always very small. However, if this population is grown in the presence of the antibiotic, the susceptible bacteria are inhibited and the resistants continue to multiply. In the end, the majority of the population will be resistant bacteria. It is said that the antibiotic has selected the resistant forms. The appearance and the spread of bacterial populations resistant to antibiotics are the result of the combined effects of mutation and selection. It can be shown experimentally that an antibiotic acts only as a "selecting agent" and not as a "mutagenic agent." This is done by showing that the frequency of appearance of resistant bacteria in a susceptible population is independent of the presence or absence of the antibiotic, which is to say that resistant mutants can arise in a bacterial population before the population comes into contact with the antibiotic. Different investigators have used various experimental approaches to demonstrate this process. A simplified example of this type of experiment is presented and explained in Fig. 4.1.

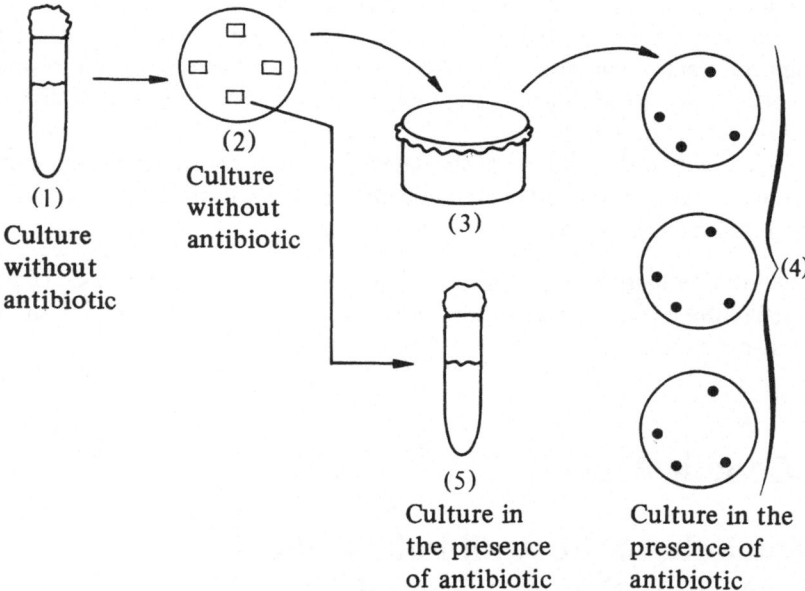

Figure 4.1. Bacteria from a culture of *E. coli* (**1**) susceptible to streptomycin are used to seed an agar plate (**2**) and are allowed to grow overnight. With the aid of a velvet plug (**3**), the plate is then replicated on a few agar plates (**4**) containing streptomycin. Here only bacteria resistant to the antibiotic can multiply and form colonies. The identical positions of these colonies on the replica plates demonstrate that the resistant bacteria were present in the original plate without antibiotic and were transferred during the replication process. In fact resistant cells can be isolated from areas of the original plate corresponding to the position of the colonies, and identified by subculture in the presence of streptomycin (**5**).

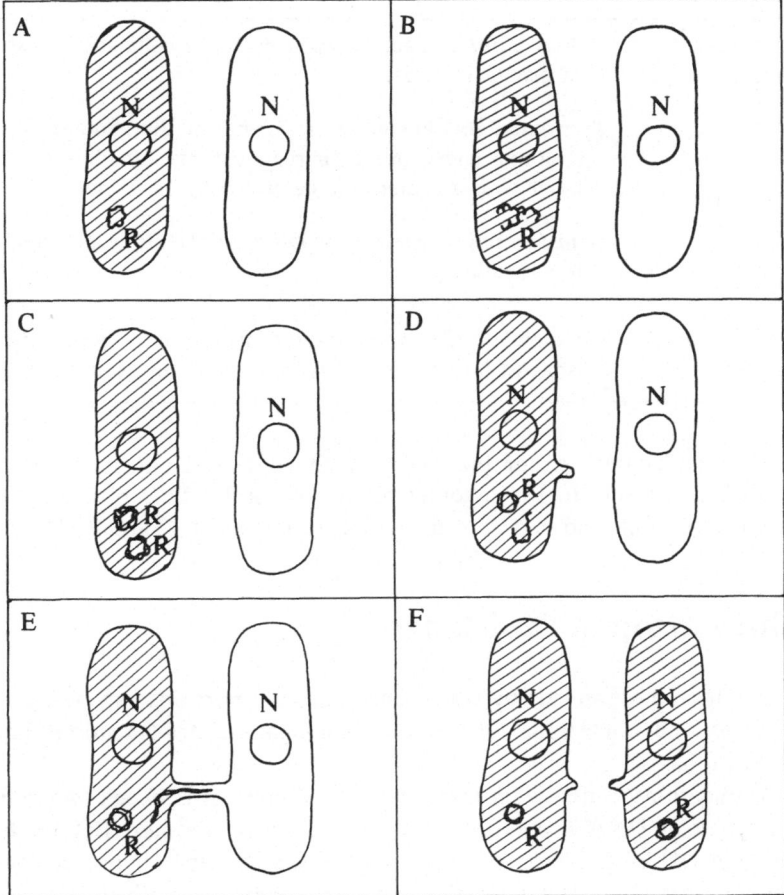

Figure 4.2. The *shaded* cell is the donor; the *unshaded* cell is the recipient. (A, B, C): Replication of R factor. In **D** one sees the formation of the "pilus" on the donor that then attaches to the recipient to physically connect the two cells (**E**). One copy of R factor is transferred through the pilus to the recipient, making it a carrier of R factor. Once the transfer is completed, the cells separate (**F**).

B. Transfer of Resistance from One Bacterium to Another

The strain that has mutated and become resistant can in certain instances transfer the information for resistance to a susceptible strain, causing it to become resistant and capable of transfering resistance to still another cell (Fig. 4.2). Genetic information for resistance can be contained within

Table 4.2. Mechanisms of Transfer of Resistance

Chromosomal resistance	*Conjugation:* contact between bacterial cells and transfer of genetic material
	Transduction: transfer of fragments of the genome from a donor bacterium to a recipient by infection of the latter with a bacteriophage grown in the donor
	Transformation: transfer of DNA fragments from a donor to a recipient
Extrachromosomal resistance	R-factors of Gram-negatives transferred by conjugation Plasmids of Gram-positives transferred by transduction

one of two molecules: either the DNA of the bacterial chromosome or small DNA fragments in the cytoplasm, the plasmids. This resistance is called chromosomal, and extrachromosomal, respectively (Table 4.2).

C. Chromosomal Resistance

There are three mechanisms by which chromosomal portions can be transferred from one bacterium to another: conjugation, transduction, and transformation.

1. Conjugation occurs when there is physical contact between two cells and a portion of DNA passes from one bacterium to another. Before the transfer, the DNA of the donor bacterium replicates. Once it has entered the recipient cell, the DNA from the donor can exchange with the homologous piece in the recipient cell's DNA. If the piece transferred contains the genetic determinant for resistance to an antibiotic, both the donor and recipient bacteria will be resistant.

2. Transduction refers to the transfer of a piece of DNA from one bacterium to many others through a bacteriophage. The phage infects the bacterium, takes up a piece of bacterial DNA, and incorporates it into its own DNA. It then replicates (the replica containing the piece of bacterial chromosome as well), lyses the host cell, and infects other bacteria. If the incorporated piece of DNA happens to contain the genetic determinant for resistance to an antibiotic, the result of the transduction will be transmission of resistance from one to many bacteria.

3. Transformation consists of the liberation into the environment of DNA from a bacterium and its uptake into another. The role of this phenomenon in transfer of resistance between bacteria is not very well known, and appears to be of marginal importance.

D. Extrachromosomal Resistance

In addition to the chromosome, the bacteria may contain cytoplasmic pieces of genetic material (DNA), called plasmids, that give it certain characteristics.

1. R Factors

One type of plasmid, called R factor, confers the property of resistance to one or more antibiotics on the bacteria that contain it.

a. Demonstration of the Existence of R Factor

Some epidemiological and laboratory data gave rise to the suspicion that there must exist an extrachromosomal factor for multiple resistance capable of transferring resistance to other bacteria. In 1959, studies of an epidemic of intestinal shigellosis (bacillary dysentary) in Japan confirmed this suspicion. After the antibiotic treatment, the patients who had first excreted susceptible bacteria in the stools began to excrete bacteria resistant to four different antibiotics (multiple resistance). It is very difficult to explain the appearance of multiple resistance by mutation. Spontaneous mutation toward resistance to a single antibiotic occurs with a frequency of about 10^{-8}. The frequency of appearance of four independent mutations conferring resistance to four different antibiotics would be $10^{-8} \times 10^{-8} \times 10^{-8} \times 10^{-8} = 10^{-32}$, a very rare event indeed. The multiresistant bacteria isolated during the Japanese epidemic were too numerous and appeared too quickly after antibiotic treatment for multiresistance to occur. In addition, why would the antibiotic given have selected multiresistant bacteria? Since bacteria resistant to only one antibiotic appear with higher frequency than multiresistants, the bacterial population should be comprised essentially of strains resistant only to the selecting antibiotic.

The multiresistant bacterial strains isolated from patients treated with the antibiotic were mixed in test tubes with susceptible bacteria, and it was found that the susceptible bacteria acquired the multiple resistance. It was known at that time that the individual chromosomal genes responsible for resistance to the four antibiotics were situated in different positions along the chromosome. Therefore, simultaneous transfer would require transfer of almost the entire chromosome. This never occurs in nature, since only small pieces of chromosome are transferred in conjugation, transduction, and transformation. Therefore, it was postulated that there must be genes for resistance to the four antibiotics that were not the same as those in the chromosome and that these genes were concentrated in small pieces of extrachromosomal DNA (plasmids) capable of being transferred from one bacterium to another. These plasmids were named R factors.

b. Physical and Genetic Nature of R Factors

The classic techniques of molecular biology were applied to the R factors. It was shown that these are double-helical DNA in the form of a covalently closed ring, with molecular weights of about 10^6–10^8 daltons.

In addition to the essential property of being able to make host cells resistant to antibiotics, the R factors have other characteristics: Multiplication of R factors does not require their incorporation into the bacterial chromosome, which means they are self-replicating units; their replication is coordinated with that of the chromosome, as shown by the fact that they are present in the cell in a 1:1 ratio with the chromosome; they are transferable; i.e., they can transfer a copy into a recipient bacterium (Fig. 4.2).

R factors are made up of (1) a factor responsible for their replication and transmission, called resistance transfer factor (RTF) and (2) several determinants for resistance. The different units can be detached and exist autonomously in the bacterial cytoplasm.

c. Distribution of R Factors

R factors are found in many species of Gram-negative bacteria, especially those of the Enterobacteriaceae (intestinal bacilli in the mammalian gastrointestinal tract). They can be transferred between bacteria of different species, e.g., between *Escherichia coli* and *Shigella*. They are present in both pathogenic and nonpathogenic organisms. They can therefore be found in the normal intestinal flora. This flora can serve as a storehouse for resistance, and transfer the R factors to pathogenic bacteria (*Shigella, Vibrio, Klebsiella, Proteus, Pasteurella, Hemophilus,* etc.).

2. Plasmids in Gram-Positive Organisms

In Gram-positive bacteria (e.g., pyogenic cocci such as staphylococci, streptococci, diplococci), there are plasmids similar to the R factors of Gram-negative organisms. However, they can be transferred from one bacterium to another by a phage, by means of processes similar to transduction. The types and mechanisms of resistance for the more common antibiotics are summarized in Table 4.1.

Chapter 5

Activities of the Antibiotics
in Relation to Their Structures

I. Introduction

This chapter describes the principal classes of antibiotics and attempts to demonstrate the relationships that exist between their chemical structures and their biological properties, especially their antibacterial activities, mechanisms of resistance, and toxicological profile. Knowledge of these relationships provides the basis for a rational search for new derivatives. Chemical modification of the antibiotics is a method for translating information obtained in this type of study into practical use, resulting in development of new products with more effective therapeutic properties.

A. Bases of the Structure–Activity Relationship

The activity of an antibiotic can be described at the molecular level in terms of its capacity to form a complex with a macromolecule (e.g., protein or nucleic acid) essential for the growth of the bacterial cell, thus rendering the macromolecule nonfunctional. Except when a covalent bond is formed between antibiotic and macromolecule, complexes are formed through so-called weak bonds between certain chemical groups of the antibiotic and of the macromolecule.

The energy in any single one of these bonds is normally not sufficient to ensure the formation of a stable complex. Therefore, if inhibition is to occur, several weak bonds must form between several functional groups of the antibiotic and the macromolecule.

The strength of a weak bond is dependent on the distance between the interacting atoms. Several bonds of sufficient strength can form between the two molecules only when the spatial structure of the antibiotic permits several functional groups to come within the correct distance for interaction.

Therefore, we could look at the antibiotic molecule as a rigid structure, a framework that carries and maintains in correct positions some functional groups able to interact with the macromolecule. It is intuitively obvious that in a very small molecule any changes in structure will cause a change in the position of the functional groups participating in the formation of the bonds, with a consequent loss of activity. We can also deduce that in a more complex molecule changes can be produced that do not affect these functional groups or their relative positions.

The study of the relationships between chemical structure and biological activity of an antibiotic consists of identifying those functional groups directly involved in the activity at the molecular level and of determining what changes can be made in the molecule without causing it to lose activity.

B. Chemical Modifications of Natural Antibiotics

Changes in the molecule that do *not* interfere with the activity at the molecular level, i.e., with the capacity to form a complex with a macromolecule in the bacterial cell, can be of fundamental importance with regard to the in vivo biological activity and the practical usefulness of the antibiotic. Properties such as transport across the membrane, and hence penetration into the bacterial cell, absorption and distribution throughout the body of an animal, and susceptibility to the actions of different enzymes can be profoundly influenced by more or less drastic modifications in the original molecule.

During the 1950s, there was considerable skepticism about the possibility of improving antibiotics substantially by semisynthetic processes, i.e., chemical modification of the natural product. This was based on the lack of success of attempts made on the products then in use and on the conviction that it would be impossible to improve the activity of a natural product.

The first argument was obviously not valid, as it implied that all antibiotics were members of one class and extrapolated to all of them a property that had been demonstrated for only a few examples. As for the belief that nature could not be improved, it is admittedly difficult to improve a product that is the result of selective evolution over millions of years. However, in the case of an antibiotic, this evolutionary selection was based in one way or another on the survival of the producing microorganism in its natural habitat, and certainly not on the therapeutic activity of the antibiotic, which is our measuring device for defining "improvement" as the result of some change in the molecule.

The results discussed in the following sections amply demonstrate the advantages that can be obtained by chemical modification of antibiotics. Today, in fact, most of the antibiotics used clinically are semisynthetic.

II. β-Lactam Antibiotics

The antibiotics of this class are chemically characterized by the presence in the molecule of a β-lactam, a cyclic amide forming a four-atom ring. The presence of this intact ring is essential for activity. The opening of this ring, which can occur rather easily either chemically or enzymatically, is both the main difficulty encountered in the chemical manipulation of the molecule and the most common cause of biological inactivation by bacterial enzymes. All β-lactam antibiotics have a common mechanism of action: inhibition of bacterial cell wall synthesis. However, differences have been shown, as discussed in Chapter 3, in their capacity to inhibit different enzymes.

The β-lactam group includes two of the most important families of antibiotics, the penicillins, in which the lactam ring is fused with a thiazolidine (a five-atom ring) and the cephalosporins, in which there is a dihydrothiazine (a six-atom ring) (see Fig. 5.1). Most of the derivatives currently used in therapy have been obtained by modification of the lateral chain at position 6 of penicillin or position 7 of cephalosporin. Active

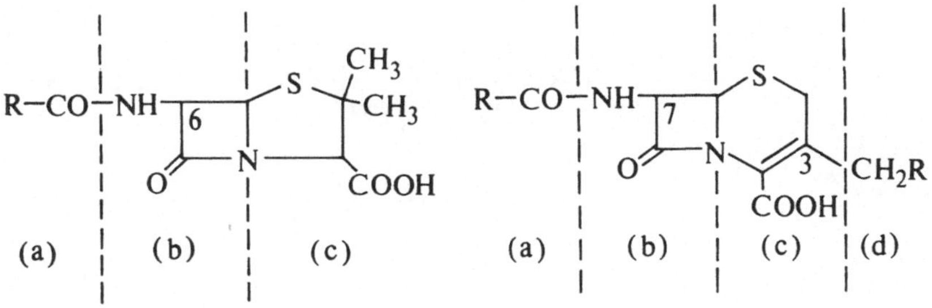

Penicillins

(a) = Side chain
(b) = β-lactam ring
(c) = Thiazolidine ring

Cephalosporins

(a) = Side chain
(b) = β-lactam ring
(c) = Dihydrothiazine ring
(d) = Group at position 3

Figure 5.1. β-Lactam antibiotics.

derivatives of the latter have also been obtained by modifications of the substituent at position 3.

In addition to these "classic" β-lactam antibiotics, several new ones recently isolated from actinomycetes are at present being intensively studied because of their interesting activities. Collectively referred to as "nonclassic" β-lactams, these include the cephamycins or 7-methoxycephalosporins and other derivatives in which the lactam is not fused with a sulfurcontaining ring.

A. Natural Penicillins

The era of extensive use of antibiotics in medicine began in 1942 when penicillin G (Fig. 5.2) was introduced into clinical practice. Penicillin G is active against Gram-positive bacteria, *Neisseria,* and *Treponema pallidum,* the causative agent of syphilis. Its great efficacy in vivo and lack of toxicity even at very high doses have made it for more than 30 years the antibiotic of choice in the treatment of several infectious diseases. The

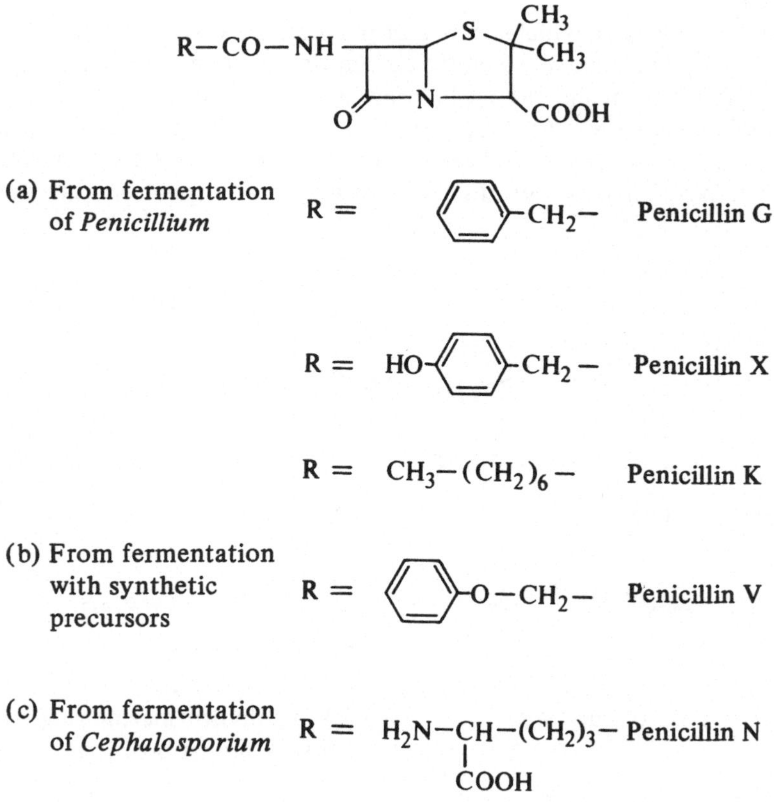

(a) From fermentation
   of *Penicillium*   R =   ⬡-CH₂-   Penicillin G

R = HO-⬡-CH₂-   Penicillin X

R = CH₃-(CH₂)₆-   Penicillin K

(b) From fermentation
   with synthetic   R =   ⬡-O-CH₂-   Penicillin V
   precursors

(c) From fermentation
   of *Cephalosporium*   R =   H₂N-CH-(CH₂)₃- Penicillin N
                                     |
                                    COOH

Figure 5.2. Some of the penicillins obtained by fermentation.

pharmacokinetic properties of penicillin G, however, are not good. It is absorbed partially when administered orally, but most of it is inactivated by the acidic pH in the stomach. When injected (as the soluble potassium salt), it is rapidly absorbed but also rapidly excreted in the urine, with a serum half-life of only 30 min. This can be circumvented by the use of insoluble salts, such as procaine penicillin, which gives blood levels for 24 h, or benzathinepenicillin, which is still detectable in serum 2 weeks after administration.

Penicillin G was chosen from among the natural penicillins produced by *Penicillium chrysogenum* (see Fig. 5.2) because: (1) Its yields in the fermentation broth could be substantially increased by the addition of the lateral chain precursor phenylacetic acid; and (2) Although penicillin K appeared to be more active in vitro, penicillin G was more effective in curing bacterial infections in animals. This revealed an important aspect of structure–activity relationships in penicillins: When the lipophilicity of the side chain is increased, the binding to the serum protein is also increased and this results in lesser efficacy.

It was soon clear that modification of the properties of penicillin G could have the following objectives:

1. improvement of absorption after oral administration;
2. enlargement of the spectrum of activity against Gram-negative bacteria;
3. reduction of the incidence of allergic reactions;
4. acquisition of activity against resistant staphylococcal strains.

The last two objectives arose from the clinical observations that (1) allergy to penicillin G is a rather common and sometimes severe side effect and (2) with the widespread use of the drug, resistant strains were rapidly emerging.

B. Penicillin Obtained by Fermentation with Synthetic Precursors

The existence of several natural active penicillins differing only in the structures of their side chains suggested that other derivatives with improved characteristics could be obtained by introducing further variations into the chain. However, for several years no chemical method was available to carry out these variations.

After the observation that the production of a given penicillin could be increased by addition to the cultures of the acid corresponding to its side chain (for example, phenylacetic acid increases the production of penicillin G), attempts were made to biosynthesize new penicillins by adding nonnatural precursors to the fermentation. Some of these were utilized by the microorganism and gave rise to new penicillins. The most important of these is penicillin V (Fig. 5.2), or phenoxymethyl penicillin, obtained after addition of phenoxyacetic acid, which is active orally since it is more resistant than penicillin G to acid degradation. However, this method had a limited potential because only a few compounds are accepted as substrates by the enzymatic system for penicillin biosynthesis.

These are a limited number of substituted phenyl or phenoxyacetic acids and some α, ω dicarboxylic acids. None of the products obtained, with the exception of the above-mentioned penicillin V, had any advantages over penicillin G.

C. 6-Aminopenicillanic Acid

The breakthrough that led to preparation of a thousand semisynthetic penicillins was the isolation of 6-aminopenicillanic acid (6-APA). This product, which from the chemical standpoint is the penicillin nucleus, was first isolated from fermentations of *Penicillium* to which no side-chain precursors had been added. Later, it was found that several microbial enzymes could split benzylpenicillin into phenylacetic acid and 6-APA. This reaction is still used industrially, although later chemical methods of hydrolysis have been described. Semisynthetic penicillins can now be prepared by addition of the appropriate acid chloride or acid anhydride to 6-APA, according to the reaction scheme shown in Fig. 5.3. Alternative methods of acylation are also available, including enzymatic ones.

D. Penicillins Absorbed after Oral Administration

Starting with 6-APA, many structural analogs of penicillin V have been synthesized. Among these, phenethicillin was the first semisynthetic penicillin used in therapy, followed by its homolog propicillin (Fig. 5.4). The

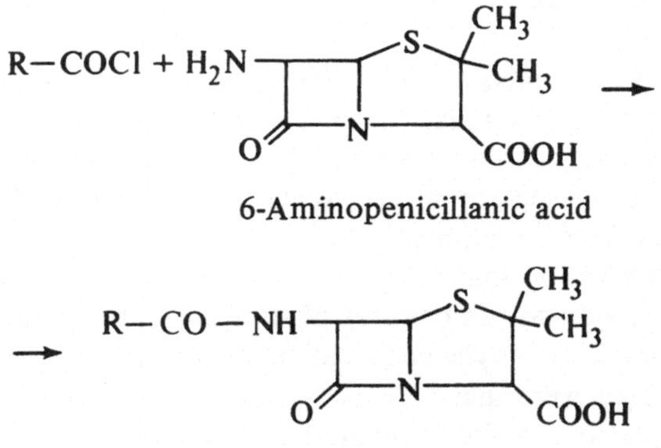

6-Aminopenicillanic acid

Figure 5.3. Scheme of synthesis of semisynthetic penicillins.

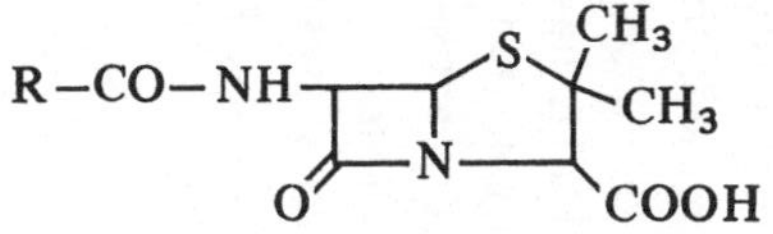

R	Name	MIC (μg/ml) *S. aureus*	*S. pyogenes*	*E. coli*	*N. gonorrhoeae*	*H. influenzae*	Oral absorption: peak serum level (% of penicillin G)
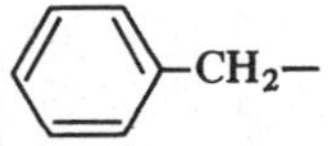	Penicillin G	0.03	0.008	64	0.008	1	100
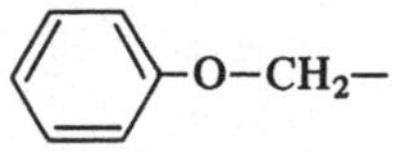	Penicillin V	0.03	0.02	128	0.03	4	250
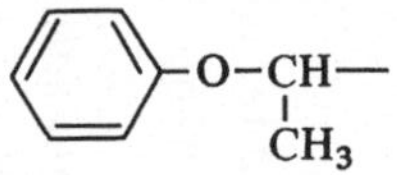	Phenethicillin	0.03	0.03	>200	0.1	4	400
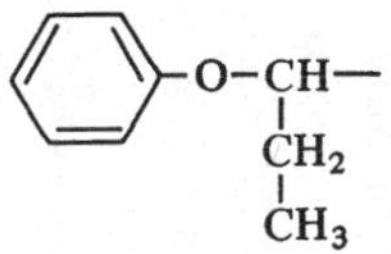	Propicillin	0.06	0.03	>200	—	—	400
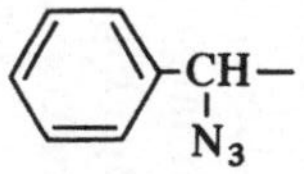	Azidocillin	0.04	0.01	—	—	0.8	300

Figure 5.4. Acid-resistant penicillins. In vitro activity and oral absorption in comparison with penicillin G.

spectrum of antimicrobial activity of these compounds is generally similar to that of penicillin G. They are, however, less active against *Neisseria* and are not recommended for the treatment of gonorrhea.

The advantage in their use is a better absorption when given orally. Two structural aspects are important for oral absorption of penicillin: (1) stability in acid medium and (2) lipophilicity of the side chain. Both experimental evidence and theoretical considerations indicate that the stability of the penicillin nucleus toward acid is influenced by the chemical nature of the chain: The presence of an electron-attracting moiety in the position α to the amide increases the stability. In penicillin V and its homologs, the electron-attracting moiety is the oxyphenyl group.

Increased lipophilicity of the chain apparently directly influences absorption from the intestine, as is true for many drugs. However, as previously mentioned, the extent of linking with serum proteins is also increased, so that the presence of long lipophilic chains results in practice in decreased efficacy. Recently, azidocillin, a new penicillin well absorbed orally, has been proposed for clinical use. Although its spectrum of activity is generally similar to that of penicillin G, azidocillin appears to be more active than the phenoxypenicillins, especially against *Hemophilus influenzae*.

Other penicillins that, in addition to being orally absorbed, possess specific properties such as activity against resistant *S. aureus* strains or an enlarged spectrum of action are discussed in the following sections.

E. Penicillins Insensitive to Staphylococcal Penicillinase

A problem that rapidly became evident with the extensive use of penicillin G was the widespread emergence of *Staphylococcus* strains resistant to this antibiotic. The resistance appeared to be due to the capacity of these strains to produce enzymes called β-lactamases or penicillinases that inactivate the penicillin by the reaction shown in Fig. 5.5.

The synthesis of a number of penicillins with structural variations in the side chain revealed that when the carbon atom α to the amide is included in an aromatic ring carrying substituents in the ortho position, the resulting steric hindrance is sufficient to protect the nearby lactam ring from the enzymatic attack. The first derivative used clinically for the treatment of infections due to penicillinase-producing staphylococci was methicillin. Because it is inactivated by acid, it must be administered by injection and in large doses (several grams per day) because of poor antibacterial activity (Fig. 5.6).

Similar properties but higher antimicrobial activity are found in nafcillin, another penicillinase-resistant penicillin used mainly in the United States. This is also given by injection since it is poorly absorbed orally. Activity against penicillinase-producing staphylococci and insensitivity to acid (and

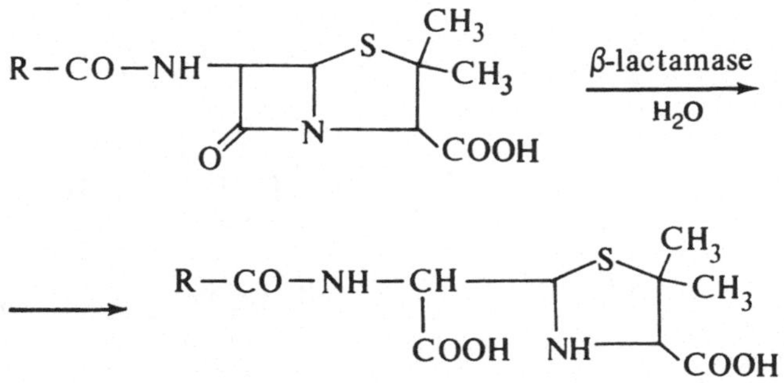

Figure 5.5. Enzymatic degradation of penicillins.

thus absorption after oral administration) are combined in isoxazolyl-penicillins. The first of these, oxacillin, is not completely satisfactory because blood levels after oral administration are rather low. Its halogen derivatives cloxacillin, dicloxacillin, and flucloxacillin give higher serum levels, not only because they are better absorbed but also because they are eliminated more slowly. The longer lasting serum levels are partially due to a greater degree of binding to serum protein. As discussed previously, this could result in reduced efficacy. Although a quantitative assessment of the significance of these different parameters is difficult, it may be noted that oxazolylpenicillins are about 10 times more active in vitro than methicillin but their therapeutic dosages are not much lower than that of methicillin.

It is important to note that with extensive clinical use, strains of *S. aureus* resistant to methicillin and to the oxazolylpenicillins have emerged. These strains are not penicillinase producers, but have a modified cell wall insensitive to the action of all known penicillins.

F. Penicillins with Enlarged Spectra of Activity

A large research effort has been devoted to the synthesis of penicillins potentially active against Gram-negative strains, such as *E. coli,* that are practically insensitive to penicillin G and to the other derivatives so far described. This research was initially directed by the observation that both penicillin N (see Fig. 5.2), a product of all cephalosporin-producing organisms and *p*-aminobenzylpenicillin, a semisynthetic derivative had some activity against Gram-negative bacteria. By comparing the activities of the many derivaties prepared, it is now possible to establish a correlation between some structural features and the property of inhibiting the growth of *E. coli* and related Gram-negative strains. This correlation can be summarized as follows:

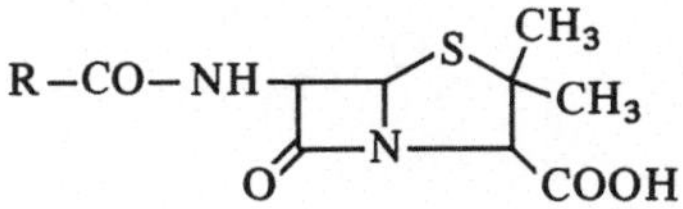

| R | Name | MIC (μg/ml) | | | Oral absorption: peak serum level (% of penicillin G) |
		S. aureus	S. aureus (penicillinase producer)	S. pyogenes	
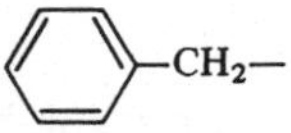	Penicillin G	0.03	125	0.008	100
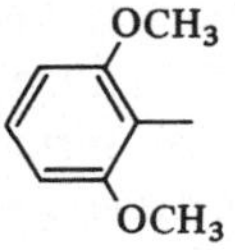	Methicillin	1	2	0.2	100
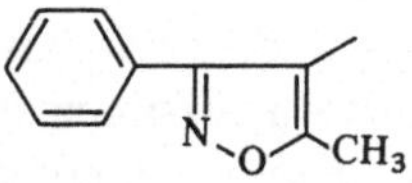	Oxacillin	0.4	0.4	0.1	200
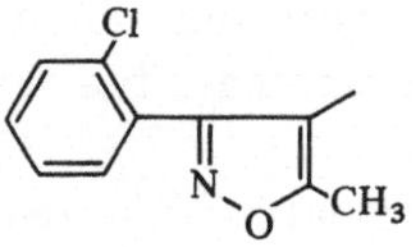	Cloxacillin	0.2	0.3	0.1	350
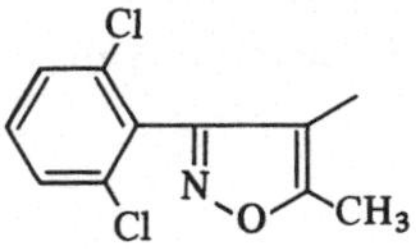	Dicloxacillin	0.06	0.1	0.05	700
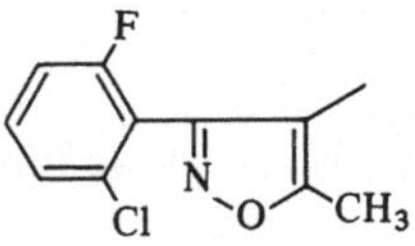	Flucloxacillin	0.1	0.3	0.05	750
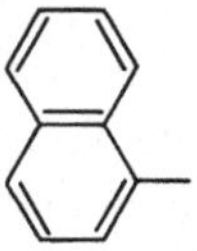	Nafcillin	0.3	0.3	0.03	100

Figure 5.6. Penicillins active against penicillinase-producing staphylococci.

1. A moderate enhancement of activity is obtained by substitution of the phenyl of penicillin G with certain heterocyclic rings.

2. The activity decreases when the lipophilic character of the chain is increased.

3. The effect of polar substituents on the chain in positions far from the amide bond is positive but small. When the substituent is an amino group this effect is more pronounced.

4. The presence of a polar group α to the amide considerably increases the activity. In this case, the stereochemical configuration is also important.

5. The activity decreases when the carbon α to the amide is fully substituted.

Among the many derivatives prepared, ampicillin (Fig. 5.7) (or D-α-aminobenzylpenicillin) was chosen for its high level of efficacy. It inhibits the bacteria sensitive to penicillin G, and most strains of *E. coli, Salmonella, Shigella,* and *Proteus mirabilis.* However, it cannot be considered an antibiotic with a broad spectrum of antibacterial activity, as it is inactive against most strains of *Klebsiella, Enterobacter,* and *Proteus* and totally inactive against *Pseudomonas.* It is sensitive to the action of staphylococcal penicillinase and thus is inactive against *S. aureus* strains that produce this enzyme.

Ampicillin is fairly resistant to acid degradation and is thus widely used orally. However, its oral absorption is not entirely satisfactory and many derivatives have been prepared to improve this aspect. The most recent of these is cyclacillin (Fig. 5.7), one of the best orally absorbed penicillins. However, cyclacillin, with a quaternary carbon in the position α to the amide, is less active than ampicillin against Gram-negative strains and is recommended mainly for Gram-positive infections. Derivatives with spectra of antibacterial activity very similar to ampicillin are epicillin and amoxycillin (Fig. 5.7). The latter is almost completely absorbed orally and thus, besides showing a higher efficacy, less frequently causes intestinal disturbances.

A different approach toward improvement of the oral absorption of ampicillin is the preparation of lipophilic derivatives that are inactive in vitro but easily hydrolyze in the body to give the active free antibiotic. Among these, the most widely used is pivampicillin (Fig. 5.7), an ampicillin ester that is well absorbed and rapidly hydrolyzed. Other esters such as talampicillin and bacampicillin have similar properties.

G. Anti-Pseudomonas and Anti-Proteus Penicillins

The increased activity against *E. coli* obtained with weakly polar groups α to the chain amide suggested the preparation of molecules with stronger polar groups in this position for their potential activity against *Pseudo-*

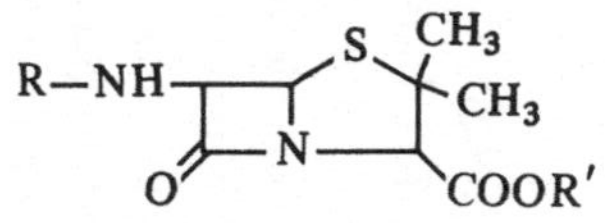

R	R'	Name	MIC (μg/ml)					Oral absorption: peak serum level (% of Penicillin G)
			S. aureus	S. pyogenes	E. coli	P. mirabilis	H. influenzae	
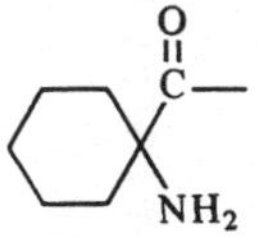 (cyclohexyl, O=C, NH₂)	H	Cyclacillin	0.3	0.2	8	4	6	750
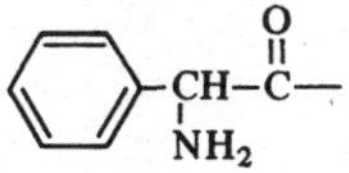 (phenyl, O=C–CH–NH₂)	H	Ampicillin	0.06	0.05	2	1.2	0.5	250
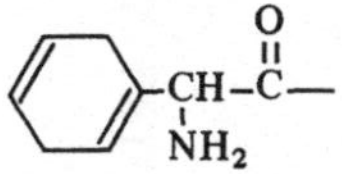 (phenyl, O=C–CH–NH₂)	H	Epicillin	0.2	0.004	1.4	1–2	0.3	250
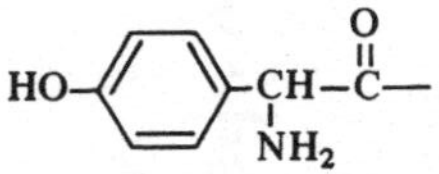 (HO-phenyl, O=C–CH–NH₂)	H	Amoxycillin	0.1	0.01	5	2–5	0.2	650
(phenyl, O=C–CH–NH₂)	–CH₂–O–C(=O)–C(CH₃)₂–CH₃	Pivampicillin	hydrolyzed to ampicillin					750

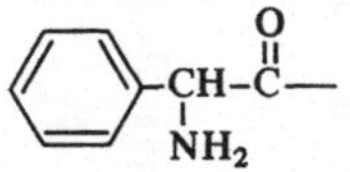

Figure 5.7. Orally absorbed penicillins, active against *E. coli* and other Enterobacteriaceae.

monas or *Proteus* strains. The attempt was successful, as products with this structural moiety are insensitive to *Pseudomonas* β-lactamase. Resistance to this enzyme action is, however, only one of the requisites for activity. Among the many products prepared, carbenicillin, sulbenicillin, and ticarcillin were sufficiently active against *Proteus, Pseudomonas,* and other Gram-negative organisms and were introduced into clinical use (Fig. 5.8). The doses needed are often very high, up to 20 or even 40 g per day in very severe cases, although much lower doses may suffice for urinary tract infections or against more sensitive organisms. These products are not absorbed orally. Esters readily hydrolyzed in the body fluids, such as carfecillin and carindacillin, are available for oral administration, but high blood levels cannot be obtained with these. More recently, a series of ureido derivatives of ampicillin, i.e., mezlocillin, azlocillin, and piperacillin (Fig. 5.8) have been successfully used in therapy by intravenous administration as anti-*Pseudomonas* and anti-*Proteus* penicillins. Their in vitro activities are quite impressive. Some doubts have been raised whether they can be used in dosage consistently lower than that of carbenicillin, because it appears that their bactericidal activity is somewhat less than their bacteriostatic activity.

H. Amidino Penicillins

The concept, based on a large amount of data, that an amide in position 6 of aminopenicillanic acid is an absolute prerequisite for activity was recently contradicted by the observation that amidine derivatives of 6-APA possess antibacterial activity. The most interesting derivative of this series is mecillinam (Fig. 5.9), which has been shown in clinical trials to be effective and was recently introduced into therapy, together with its ester pivmecillinam, which is active orally. The antibacterial activities reported in Fig. 5.9 show that mecillinam's spectrum of action substantially differs from that of the penicillins. Moreover, its mechanism of action is not identical to that of classic penicillins: It still interferes with the formation of the cell wall, but the enzyme inhibited is not the same one.

I. Modification of the Penicillin Nucleus

None of the modifications of the penicillin nucleus resulted in increased antibacterial activity. However, in some instances a good level of activity was maintained.

1. Modification of the carboxyl group: (a) Esterification brings about almost complete inactivation of products. However, as mentioned earlier, some esters are well absorbed orally and are rapidly hydrolyzed in body

Name	R	MIC (µg/ml)			
		E. coli	P. vulgaris	P. morganii	P. aeruginosa
Carbenicillin	C_6H_5–CH(COOH)–CO–	12.5	25	6	50
Sulbenicillin	C_6H_5–CH(SO_3H)–CO–	12.5	25	—	25
Ticarcillin	thienyl–CH(COOH)–CO–	5	2.5	2.5	25
Carfecillin	C_6H_5–CH(COO–C_6H_5)–CO–	hydrolyzed to carbenicillin			
Carindacillin	C_6H_5–CH(COO–indanyl)–CO–	hydrolyzed to carbenicillin			

Core structure: R–NH–(β-lactam/thiazolidine) with CH_3, CH_3, COOH substituents.

R	Name	MIC (µg/ml)			
		E. coli	P. vulgaris	P. morganii	P. aeruginosa
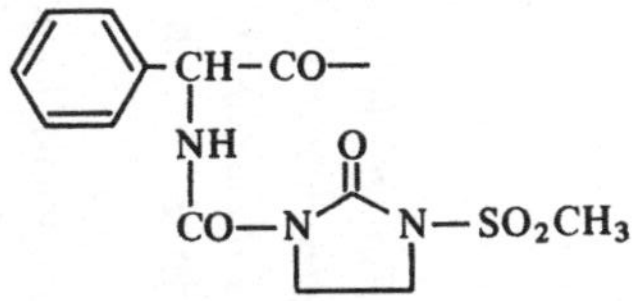 (structure with —N—SO$_2$CH$_3$)	Mezlocillin	12.5	1.5	1.5	25
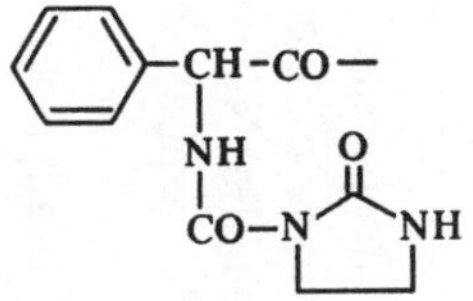 (structure with —NH)	Azlocillin	12.5	12.5	12.5	25
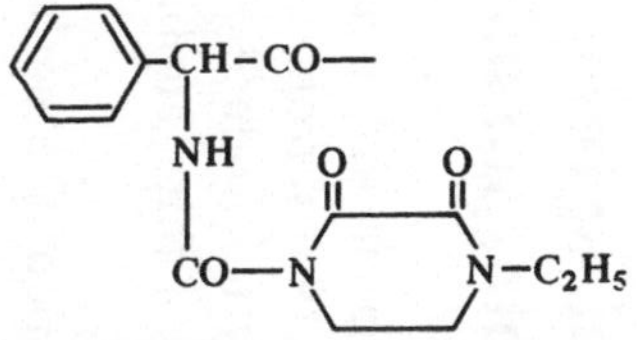 (structure with N—C$_2$H$_5$)	Piperacillin	0.8	0.8	0.8	6.2

Figure 5.8. Penicillins active against *Pseudomonas* infections.

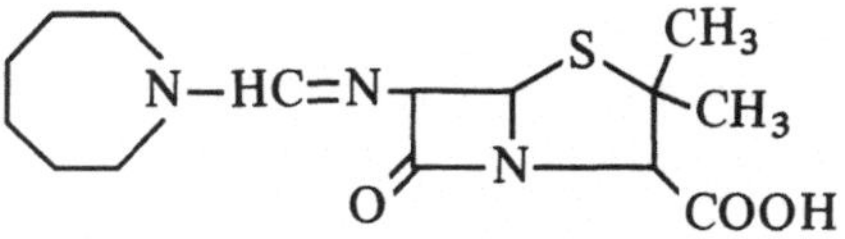

MIC (μg/ml)	
S. faecalis	>100
S. aureus	5
E. coli	0.02
H. influenzae	16
K. pneumoniae	0.1
S. typhimurium	0.1

Figure 5.9. Mecillinam: in vitro activity.

fluids, thus releasing the active penicillin. (b) Activity is retained after conversion of the carboxyl to a thioacid or to an amide. (c) Reduction to a hydroxyl group results in inactivation.

2. Structure of the nucleus: (a) Opening of either of the rings causes inactivation. (b) Inversion of configuration at one of the three centers of asymmetry causes a great reduction in activity.

3. Other substitutions: (a) The presence of the two methyl groups is not required for activity. (b) Substituting a methyl or a methoxy group for the hydrogen on carbon 6 results in decreased activity. If the group is heavier, activity disappears. (c) Oxidation of the sulfur atom to sulfoxide reduces the activity.

J. Natural Cephalosporins and 7-Aminocephalosporanic Acid

All cephalosporins originate from cephalosporin C (Fig. 5.10), an antibiotic isolated in the 1950s from cultures of *Cephalosporium,* a mold studied for its capacity to produce penicillin N (see Fig. 5.2). *Cephalosporium* also produces a third antibiotic, cephalosporin P, which has no therapeutic importance.

Not very effective, poorly absorbed orally, cephalosporin C attracted the attention of researchers because, although structurally related to the penicillins, it was active against penicillinase-producing *S. aureus* and was more active against Gram-negative bacteria. It was felt to be an interesting starting material for the preparation of semisynthetic derivatives. The experience with penicillins directed the research toward the preparation of 7-aminocephalosporanic acid (7-ACA) (Fig. 5.10), obtained from

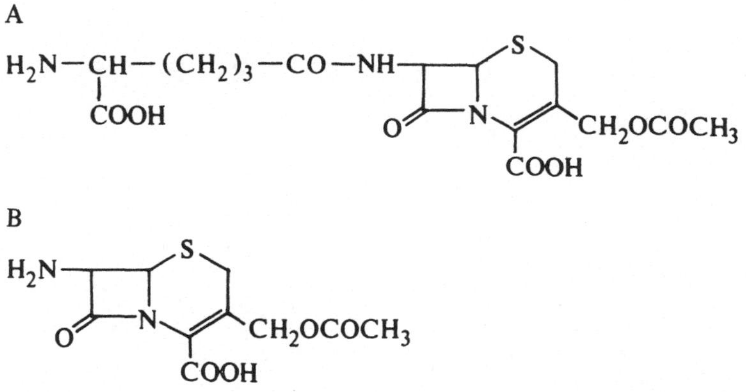

A

H$_2$N—CH—(CH$_2$)$_3$—CO—NH

|
COOH

S

O   N
CH$_2$OCOCH$_3$

COOH

B

H$_2$N

S

O   N
CH$_2$OCOCH$_3$

COOH

Figure 5.10. (A) Cephalosporin C. (B) 7-Aminocephalosporanic acid.

cephalosporin C by a rather complex chemical reaction. In contrast with penicillins, an enzymatic method to hydrolyze the amide in the chain of cephalosporin C became available only recently. This is apparently due to the structure of the lateral chain, as semisynthetic cephalosporins with different chains are easily hydrolyzed enzymatically.

With 7-ACA as starting material, a large research effort has made an enormous number of semisynthetic cephalosporins available, several of which are now in clinical use.

Cephalosporins, unlike penicillins, can be modified not only at the side chains but also in position 3 of the nucleus, where the acetoxy group can be easily eliminated or substituted without loss of activity.

K. First-Generation Cephalosporins

The first cephalosporin introduced into medical practice was cephalothin, in which a thienylacetic acid is substituted for aminoadipic acid in the side chain (Fig. 5.11). Cephalothin is active against staphylococci both sensitive and resistant to penicillin, and against *Neisseria* and most *E. coli*, *Salmonella*, and *Proteus mirabilis* strains. In infections due to these strains, its clinical effectiveness has been demonstrated.

The limitations to the use of cephalothin, which have in a way provided an impetus for further research in cephalosporins, may be summarized as follows:

1. It is not absorbed orally.
2. Intramuscular injections are painful and intravenous administration may give phlebitis.

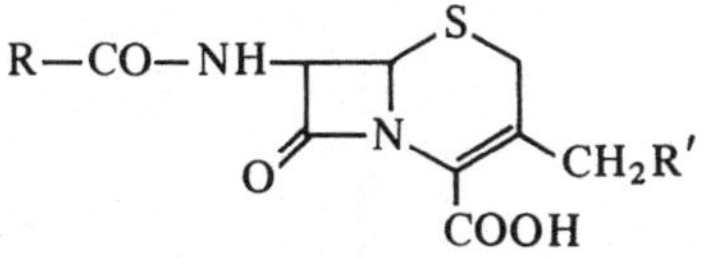

R—CO—NH

(cephalosporin nucleus with S ring, N, O, =CH₂R', COOH)

structure: R—CO—NH— attached to the β-lactam–dihydrothiazine bicyclic nucleus bearing CH_2R' and $COOH$

R	R'	Name	MIC ($\mu g/ml$)					
			S. aureus	S. aureus (penicillinase producer)	E. coli	K. pneumoniae	P. mirabilis	P. vulgaris
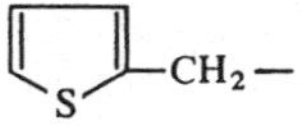 thiophene–CH_2—	—$OCOCH_3$	Cephalothin	0.2–0.4	0.4	6.2	3.2	3.2	>100
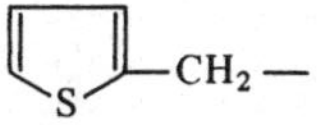 thiophene–CH_2—	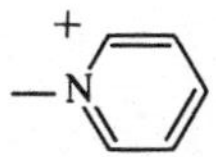 —$\overset{+}{N}$ pyridinium	Cephaloridine	0.02	0.1	3.2–6.4	3.2	6.4	50
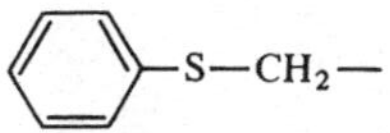 phenyl–S–CH_2—	—OCO—CH_3	Cephapirin	0.1–0.4	0.4	12	1.6	12	>100
$N\equiv C$—CH_2—	—OCO—CH_3	Cephacetrile	0.4	1.6	3.2	12	12	50
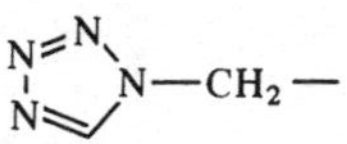 tetrazole–N–CH_2—	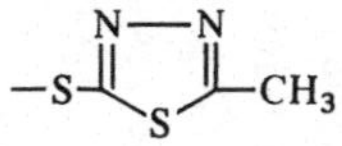 thiadiazole–S–CH_3	Cefazolin	0.2	0.8	3.2	3.2	6.4	>100

Figure 5.11. First-generation cephalosporins.

3. It is inactive against *Pseudomonas aeruginosa,* indole-positive *Proteus* species, *Serratia marcescens, Enterobacter* species, and *Bacteroides fragilis.*
4. It is rapidly eliminated (it is practically undetectable in serum 4 h after administration).

A wider range of structural modification appears to be compatible with good antimicrobial activity in cephalosporins than in penicillins. It is also more difficult to rationalize the relationship between the structure of the chemical substituents and the biological properties of the cephalosporins, in part because there are two sites in the molecule suitable for modification and the biological properties are the results of the combined effects rather that of the sum of the substitutions. Since the acetyl group in position 3 is readily hydrolyzed in the body to yield the less active hydroxyl derivatives, substitutions at this position are made to improve the pharmacokinetic properties (see Fig. 5.11).

The substitution of the acetoxy group of cephalothin with a pyridine ring produces cephaloridine. The antibacterial activity of this derivative is similar to that of cephalothin. It was introduced into clinical use because injections of cephaloridine are less painful and it gives higher serum levels. However, high doses may produce damage to kidneys, a very rare effect of other cephalosporins.

Pharmacokinetic properties similar to those of cephalothin are reported for cephapirin and cephacetrile, both of which bear a modified amide chain. Again, a better tolerability after intramuscular injection is an important rationale for its use. Cephapirin seems less active on Gram-negative bacteria, whereas cephacetrile is somewhat more active than cephalothin on *Escherichia coli.*

Cefazolin, with a mercapto triadiazole at position 3 and a tetrazole in the amide chain, has about the same activity as cephalothin on Gram-positive organisms (although it is slightly less resistant to *S. aureus* β-lactamases) but is somewhat more active against *E. coli, K. pneumoniae,* and *Salmonella.* It is better tolerated after intramuscular injections and can be administered at longer intervals, having a serum half-life of almost 3 hr.

L. Orally Active Cephalosporins

The synthesis of new cephalosporins obviously took advantage of the knowledge accumulated during years of penicillin studies. The phenylglycine side chain of ampicillin was shown to be an excellent moiety for cephalosporins also. All the orally active cephalosporins used at present, and most of those in clinical trials (see Fig. 5.12), possess either this side chain or small variants of it, such as cephradine, an analog of epicillin, and cephadroxil and cephatrizine, analog of amoxycillin. The first orally active cephalo-

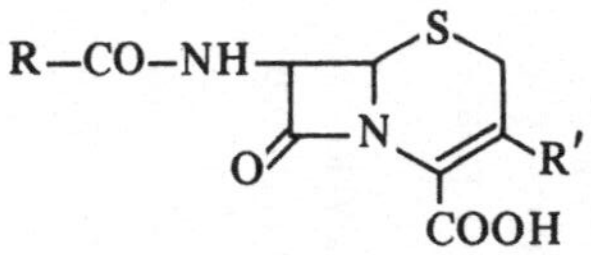

R	R'	Name	MIC (μg/ml)				Peak serum level: in men after 500 mg os (μg/ml)	Serum half-life (hours)
			S. aureus	E. coli	H. influenzae	P. mirabilis		
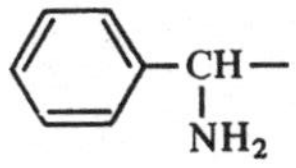 phenyl–CH(NH₂)–	–CH₂OCOCH₃	Cephaloglycin	1.6	3.2	—	6.2	1	—
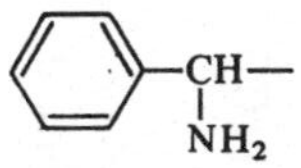 phenyl–CH(NH₂)–	–CH₃	Cephalexin	1.6–6.4	12.5	12.5	50	20	0.6–0.9
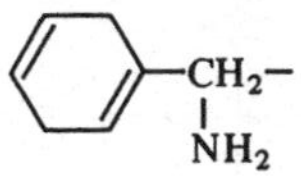 cyclohexadienyl–CH₂(NH₂)–	–CH₃	Cephradine	6.4	12.5	6.4	100	15	—
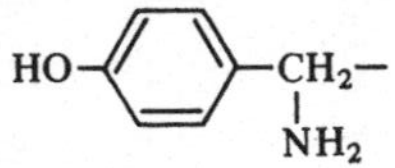 HO-phenyl–CH₂(NH₂)–	–CH₃	Cephadroxil	1.6–6.4	12.5	6.4	25	20	1.2
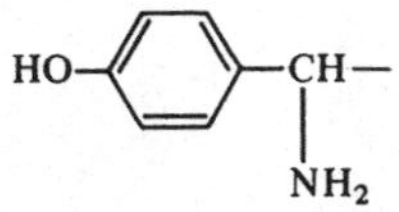 HO-phenyl–CH(NH₂)–	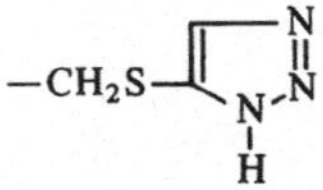 –CH₂S–(triazole, N=N / N–H)	Cephatrizine	0.4–3.2	3.2	3.2	1.6	6	2.4
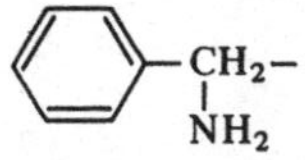 phenyl–CH₂(NH₂)–	–Cl	Cefaclor	1.6–6.4	6.4	3.2	6.4	15	0.6–0.8

Figure 5.12. Orally active cephalosporins.

sporin, cephaloglycin, had the natural substituent, acetoxymethyl, at position 3. Its spectrum of antibacterial activity is similar to that of cefoxitin, and include most of the Gram-positive bacteria (with the exception, common to most cephalosporins, of *Streptococcus faecalis*) and *Neisseria, E. coli,* and *Proteus mirabilis.* However, in general higher concentrations are needed for inhibition. At present it has been practically abandoned because it is rapidly metabolized to the poorly active desacetyl derivative. Cephalexin is more stable and better absorbed, which compensates amply for its lower in vitro activity. However, it is not recommended for treatment of serious infections but rather as "follow-up therapy." Cephradine has pharmacological and microbiological properties similar to those of cephalexin; however, it is somewhat less active on *Proteus, Klebsiella,* and *S. aureus.* Cephadroxil is also similar to cephalexin in its spectrum of antibacterial action. It has been recently introduced in clinical practice because it shows a longer serum half-life.

A combination of good absorption and greater activity against some Gram-negative bacteria is sought in the newer derivatives, mainly by varying the substituent at position 3. Cephatrizine belongs to this type of cephalosporin. Its absorption after oral administration is in fact rather poor; however, it is generally more active than the previously available derivatives on several strains including *Hemophilus influenzae.* An important achievement appears to be cefaclor, which is at least equivalent to cephalexin in activity against Gram-positive organisms and is more active against Gram-negatives, including *H. influenzae* and *N. gonorrhoeae.* It appears thus to be an useful addition to available oral cephalosporins.

M. Second- and Third-Generation Injectable Cephalosporins

The second- and third-generation cephalosporins are characterized by their enhanced antibacterial activity, obtained mainly by the choice of a suitable substituent at the amide chain in position 7 combined with a substituent in position 3 compatible with good pharmacokinetics (see Fig. 5.13).

The second-generation cefuroxime and cefamandole antibacterial spectra include indole-positive *Proteus* species, *Enterobacter,* and *H. influenzae.* Cefamandole maintains good activity against Gram-positive species and is more active than first-generation cephalosporins on *E. coli, K. pneumoniae,* and *P. mirabilis.* The third-generation cephalosporins are generally very active on "difficult" Gram-negative strains but somewhat less active on Gram-positives such as *S. aureus* and *Streptococcus.* Typical examples are cefoperazone, active on *Enterobacter* and indole-positive *Proteus* and fairly active on *Pseudomonas aeruginosa,* and cefotaxime, whose structure is a

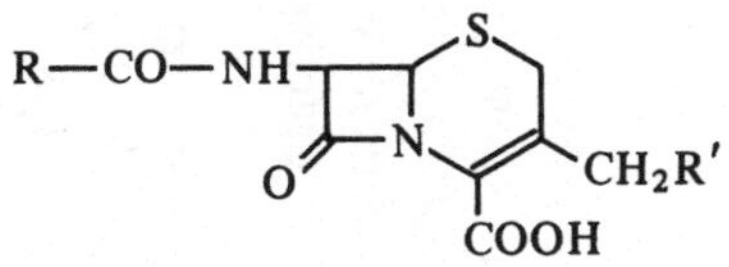

Name	MIC (μg/ml)					
	S. aureus	S. pyogenes	E. coli	P. mirabilis	P. vulgaris	P. aeruginosa
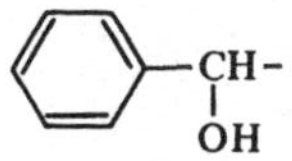 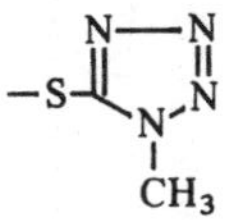 Cefamandole	0.2–0.8	0.04–0.1	0.4–0.8	1.6	25	>100
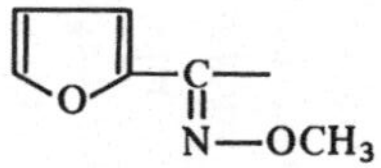 Cefuroxime	0.8	0.01	0.8–6.2	0.4	25	>100
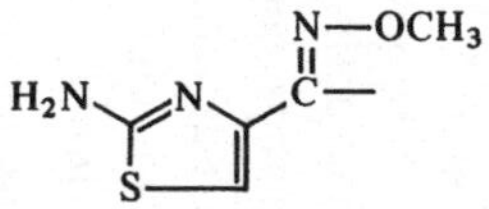 Cefotaxime	3.2	0.02–0.04	0.04–0.1	0.006	0.1	12.4–50
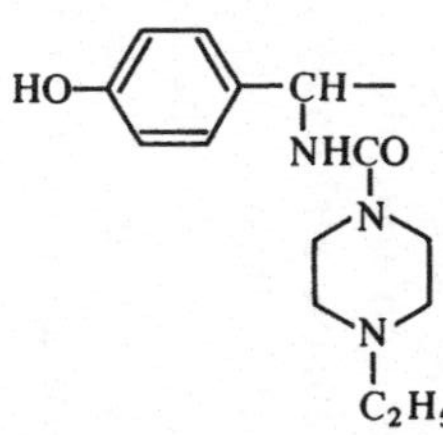 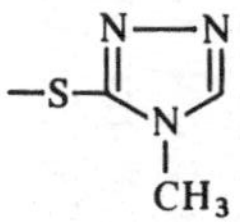 Cefoperazone	1.6	0.1–0.2	0.04–0.4	0.2	0.4	6.2–12.4

Figure 5.13. Second- and third-generation cephalosporins.

clear example of how a substituent α to the amide can, by steric hindrance, protects the molecule from the attack of inactivating enzymes.

N. Nonclassic β-Lactams

1. Cephalosporins Modified in the Thiazine Ring

The structural requirements for activity in cephalosporins are somewhat less stringent than for penicillins. Although opening of the rings constituting the nucleus of the molecule or saturating the double bond results in complete inactivation, other modification are not incompatible with activity. For example, introduction of a methoxy group at position 7, as illustrated in the next paragraph, may confer useful properties.

An observation that may prove of major importance is that in cephalosporins the ring sulfur atom can be sustituted with an oxygen or a methylene without substantial loss of activity. This opened the way to the synthesis of completely new series of derivatives. One of these, moxalactam (Fig. 5.14), shows an antibacterial activity comparable to that of third-generation cephalosporins. It is not very active on Gram-positives but it is active on *E. coli, Proteus* strains both indole-positive and -negative, *Serratia* species, and *Bacteroides fragilis*. In addition it shows some activity against *P. aeruginosa*.

2. Cephamycins

Cephamycins are naturally occurring β-lactam antibiotics isolated first in 1971 from fermentations of *Streptomyces* species. They are closely related to cephalosporins, from which they differ in the presence of a methoxy group at carbon 7.

Among the several natural ones cephamycin C appeared to have the greater activity, particularly against Gram-negatvie bacteria. The presence of the 7-methoxy group in fact may protect the molecule against some Gram-negative β-lactamases. Like cephalosporin C, cephamycin C was thus considered a suitable starting material for the synthesis of derivatives. Two of these, cefoxitin and cefmetazole (Fig. 5.14), are at present used in several countries. Their activities are similar to those of second-generation cephalosporins. They are mainly recommended for treating infections by Gram-negatives, especially *Proteus* and *Bacteroides*.

3. Recently Discovered Nonclassic β-Lactams

From the fermentation broths of various *Streptomyces* species, several β-lactam antibiotics have recently been isolated whose structures differ substantially from those of penicillin or the cephalosporins. Some of these

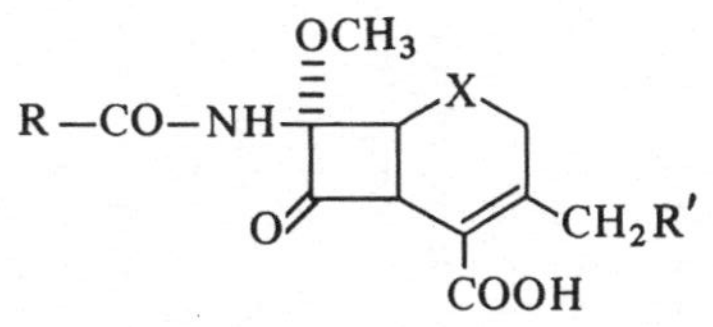

X	R	R'	Name	MIC (μg/ml)				
				S. aureus	E. coli	P. mirabilis	P.vulgaris	P. aeruginosa
$-O-$	HO–⟨⟩–$\overset{\underset{\text{COOH}}{\|}}{\text{CH}}$–	$-S-\overset{N=N}{\underset{N-CH_3}{\|\quad\|}}$	Moxalactam	3.2	0.1–0.2	0.2	0.4	25
$-S-$	$H_2N-\overset{\underset{\text{COOH}}{\|}}{CH}-(CH_2)_3-$	$-O-CO-NH_2$	Cephamycin C	160	16	—	—	—
$-S-$	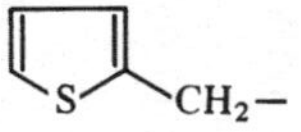 thiophene–CH_2-	$-O-CO-NH_2$	Cefoxitin	1.6–3.2	1.6–6.2	3.2	3.2	>100
$-S-$	$N\equiv C-CH_2-S-CH_2-$	$-S-\overset{N=N}{\underset{N-CH_3}{\|\quad\|}}$	Cefmetazole	0.8–1.6	1.6	3.2	1.6	>100

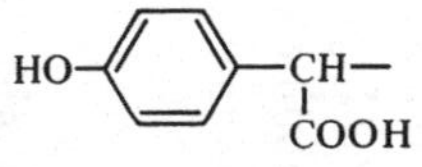

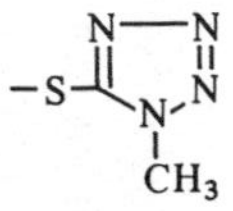

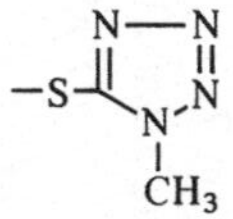

Figure 5.14. Cephamycins and moxalactam.

products are of interest for their antibacterial activities, which include activity against strains generally insensitive to classic β-lactams; others inhibit some β-lactamases and thus have synergistic activity when combined with penicillins or cephalosporins.

The activities of some typical representatives of this class of compounds are presented in Fig. 5.15. The development of these products and of their analogs is still in a relatively early stage, and it is impossible at present to predict which compound will reach clinical use. Thienamycin can be considered as a representative of the new β-lactams with high antibacterial activity, whereas clavulanic acid is a typical β-lactamase inhibitor. Augmentin, the combination of amoxycillin and clavulanic acid, for instance, shows good effectiveness against *Proteus* and *Klebsiella.*

Recently, monocyclic β-lactams have been isolated from fermentation broths of *Nocardia* spp. (nocardicins) and *Chromobacterium violaceum* and *Acinetobacter* (monobactams). Although of novel structure, monobactams have been shown to have a mode of action similar to that of cephalosporins, affecting specifically septation.

O. Unresolved Problems in the Use of β-Lactam Antibiotics

In spite of a vast amount of progress, limitations and problems still exist in the clinical use of β-lactam antibiotics concerning the spectrum of activity, the emergence of resistant strains, and tolerability.

1. Spectrum of Activity

Although no single penicillin or cephalosporin can be considered a truly broad-spectrum antibiotic, as a class they are active against almost all the bacteria responsible for the most common infectious diseases except mycobacteria and, obviously, the mycoplasma and other organisms that lack a cell wall. However, in addition to the problem of resistance, discussed later, the level of activity against several pathogens is unsatisfactory. For example, among Gram-positive organisms, *Streptococcus faecalis* is only moderately sensitive to penicillins and almost totally insensitive to most cephalosporins. None of the established classic β-lactams had a reliable activity against *Enterobacter, Klebsiella, Serratia,* or indole-positive *Proteus.*

Some cephalosporins and cephamycins recently made available show increased effectiveness against infections caused by these bacteria, but high dosages are still needed and unsusceptible strains are often encountered. Similarly, treatment of *Pseudomonas aeruginosa* infections requires very high doses of even the most active penicillins. An enormous research effort has been devoted to the synthesis and evaluation of new β-lactam antibiotics. Promising derivatives are at present undergoing clinical trials.

Name	MIC (μg/ml)		
	S. aureus	*E. coli*	*P. vulgaris*
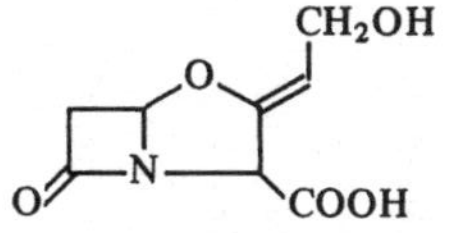 Clavulanic acid	7	31	31
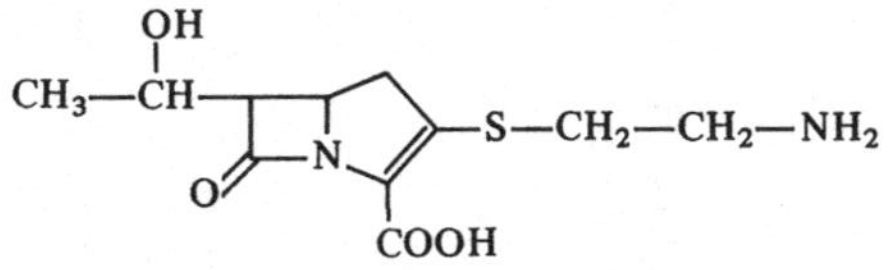 Thienamycin	0.02	0.3	5
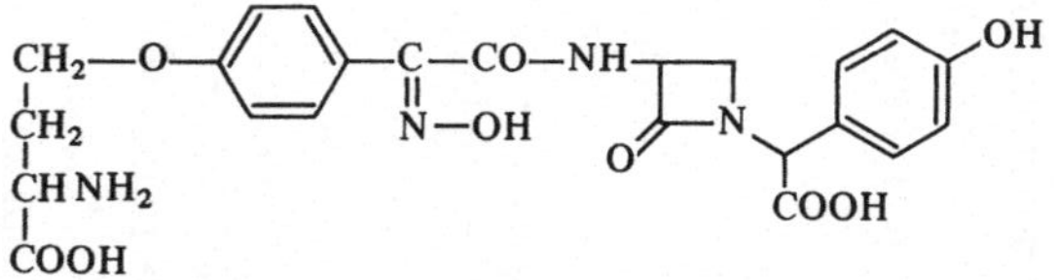 Nocardicin A	>800	100	3
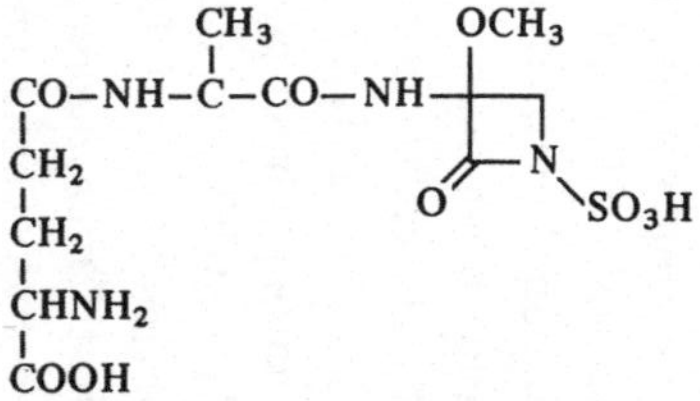 Sulfazecin	200	12	6

Figure 5.15. Nonclassic β-lactams under evaluation.

2. Resistant Strains

The most frequent mechanism of resistance to β-lactam antibiotics is the production of enzymes, β-lactamases, that inactivate the antibiotic by opening the lactam ring. Different genera of microorganisms produce β-lactamases that differ in their substrate specificities. Generally speaking, (1) β-lactamases produced by *Staphylococcus aureus, Bacillus* and *Klebsiella* inactivate penicillins only. (2) Those produced by *Enterobacter, Proteus* (indole-positive), and *Serratia* specifically inactivate the cephalosporins. (3) Some β-lactamases have a broader activity; those produced by some *E. coli* strains inactivate both cephalosporin and penicillin. Sometimes this wider activity is due to the presence of two different enzymes produced by the same strain.

In several cases, the genetic information for the production of β-lactamases is carried by plasmids or episomes and can be transferred from cell to cell, which explains the diffusion of the resistance.

The structural requirement for insensitivity to β-lactamase inactivation, i.e., a bulky substituent in the α position of the acyl side chain, has been previously discussed. Unfortunately, in penicillins this structural feature is incompatible with good activity against Gram-negative strains. Hopes for the future reside in the newer cephalosporins and in some nonclassic β-lactams that seem to be excellent inhibitors of β-lactamases. It is important to note that other mechanisms of resistance are emerging, in addition to the production of inactivating enzymes. The so-called "intrinsic resistance" is now fairly frequent, even in genera such as *S. pneumoniae,* considered in the past to be consistently sensitive. Another phenomenon becoming clinically important is that of "tolerance." Tolerant strains are inhibited by a "normal" MIC but show a much higher MBC than nontolerant strains. Penicillins behave as bacteriostatic rather than as bactericidal antibiotics toward these strains. Since with the usual dosages of penicillins the effectiveness is linked to their bactericidal action, the infection may not be cured.

3. Tolerability and Hypersensitivity Reactions

Penicillins are remarkably well tolerated antibiotics, even at very high dosages, but often cause hypersensitivity reactions that may take different forms: (1) an immediate anaphylactic reaction, which is very rare but may be lethal; (2) serum sickness, with fever, malaise, and joint pain; (3) rashes and urticaria. The frequency of hypersensitivity seems to be lower with the semisynthetic derivatives than with penicillin G, although ampicillin very frequently causes rashes. There is no agreement among investigators about the cross-sensitivity among penicillins and cephalosporins, but it is considered advisable not to treat with cephalosporins patients whose immunological sensitivity to penicillins is known.

The molecule of penicillin is too small to be immunogenic per se. Its capacity to induce antibodies may be due to: (1) the penicillin molecule forming a covalent linkage (e.g., through the opening of the lactam ring) with a host protein; (2) an impurity contained in the penicillin preparation combining with a protein; (3) penicillin preparations containing a polymer of penicillin large enough to be immunogenic. It has been demonstrated that none of these hypotheses can be discarded and probably all these causes are relevant. However, it is not possible to assess how much each mechanism contributes to the frequency of the clinical observations.

As stated previously, penicillins and cephalosporins, except in cases of hypersensitivity reactions, are well tolerated. Specific side effects are a high incidence of intestinal distress (diarrhea) with oral cephalosporins. Phlebitis is rather common after intravenous administration of most cephalosporins. The incidence of renal toxicity seems low except with cephaloridine. Some broad-spectrum derivatives, such as cefamandole, may cause bleeding in debilitated patients, possibly because of elimination from intestine of vitamin K–producing microorganisms

III. Tetracyclines

Until a few years ago, the tetracyclines were considered to be the most important antibiotics for clinical use. They are still in use, in spite of the introduction of the new cephalosporins and aminoglycosides. Nevertheless, many aspects of the biological properties of the tetracyclines have not been completely clarified. They are antibiotics with a broad spectrum of activity, are bacteriostatic, and act by inhibiting protein synthesis. The mechanism by which they block ribosomal function is not known. In theory they should be toxic substances, since in vitro they also inhibit the ribosome of eukaryotic cells, but in fact they are well tolerated. The specificity of the toxicity of tetracyclines for bacteria resides in their ability to accumulate it. It is possible that this property, and perhaps the activity of the tetracyclines, is connected with their capacity to chelate bivalent ions such as calcium and magnesium.

A. Natural Tetracyclines

The first tetracycline used in therapy, in 1948, was chlortetracycline, produced by *Streptomyces aureofaciens*. This was followed 2 years later by oxytetracycline, isolated from *Streptomyces rimosus*, and in 1952 by tetracycline (Fig. 5.16). The latter was originally synthesized by catalytic hydrogenation of chlortetracycline and later by fermentation first by growth

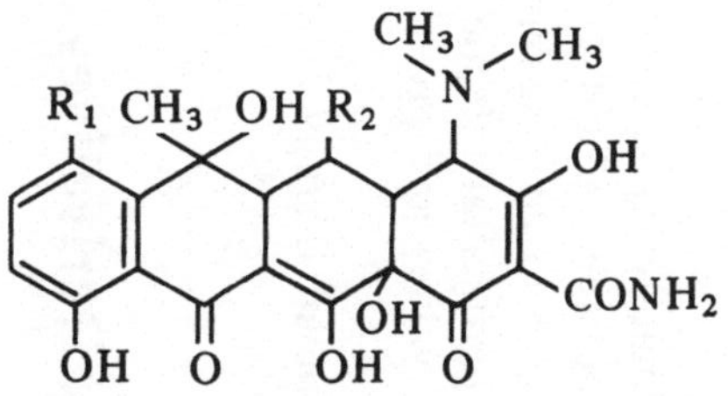

R_1	R_2	Name	S. aureus	S. pneu-moniae	S. fae-calis	E. coli	K. pneu-moniae	P. vul-garis	S. tiphy-murium	P. aeru-ginosa
						MIC (μg/ml)				
H	H	Tetracycline	0.4	0.15	0.7	1.2	0.6	4	2.3	25
Cl	H	Chlortetracycline	0.3	0.1	0.4	1.4	0.3	4.6	1.2	14
H	OH	Oxytetracycline	0.6	0.3	2	1.2	0.6	3.1	1.6	25

Figure 5.16. Natural tetracyclines.

of *Streptomyces aureofaciens* in a medium without chlorine and later by fermentation of *Streptomyces* strains that were able to produce it directly.

All three of these compounds demonstrate complete cross-resistance and all have very similar spectra of activity that are particularly broad, encompassing not only Gram-positive and Gram-negative bacteria but also such intracellular microorganisms as *Rickettsia* and mycoplasma (which are not susceptible to penicillin and cephalosporin because they have no cell walls). But they have very little activity against *Proteus* and *Pseudomonas*. In very high concentrations they also inhibit some protozoa. In spite of some claims, they have no effect against viruses, which is only to be expected from their mechanism of action.

The therapeutic effectiveness is approximately equal for all the tetracyclines and all are well, though not completely, adsorbed when given orally. Chlortetracycline is slightly more active in vitro against some bacteria, but does not give better results in experimental infections, probably because it is unstable at physiological pH values and gives lower blood levels.

The compound in use at present for human therapy is almost exclusively tetracycline.

B. Semisynthetic Tetracyclines: Derivatives at the Amide Group

The first attempts to chemically modify the tetracyclines had a very practical and limited objective: to obtain a derivative that would be soluble in water at neutral pH and could therefore be injected without causing pain or irritation. This objective was rapidly reached by synthesis of rolytetracycline (Fig. 5.17) and of the analogous compounds limecycline and mepicycline, obtained from tetracycline by reaction with formaldehyde and the appropriate amines. These products have the same activity as tetracycline and, in fact, are hydrolyzed to it in aqueous solution. Other amide derivatives that are not easily hydrolyzable are inactive.

C. Tetracycline Produced Biosynthetically

The natural tetracyclines are rapidly converted into inactive derivatives by treatment with acid or base. Degradation products, made for the purpose of determining the structure, are also inactive. Therefore, it is not surprising that the first active derivatives, apart from the water-soluble ones mentioned previously, were obtained by modifying the biosynthetic pathways.

The first interesting example was 7-bromotetracycline, which was ob-

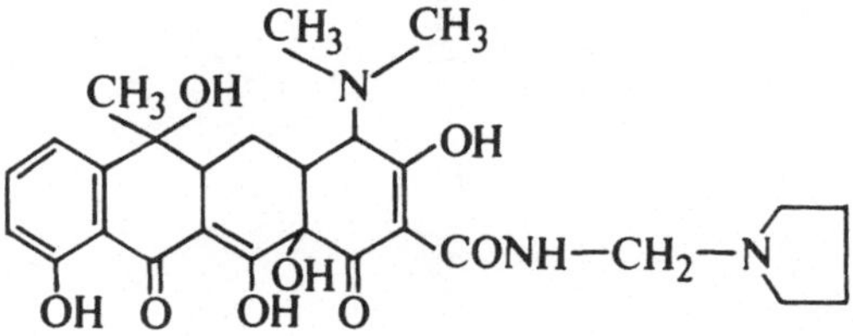

Figure 5.17. Rolytetracycline, a soluble derivative of tetracycline.

tained when bromide ions were substituted for chloride ions in the fermentation broth of *S. aureofaciens*. However, this product was less active than the natural tetracyclines and was never used.

A product that has been introduced into clinical therapy is 6-demethyl-7-chlortetracycline (Fig. 5.18), which is obtained by adding inhibitors of methylation, such as sulfonamides, to the fermentation broth or by isolating mutant strains with an inactive methylating enzyme, thereby producing the demethylated derivative instead of chlortetracycline. Demethyltetracycline and demethyloxytetracycline can be obtained by the same methods. Demethylchlortetracycline is very stable and has an antibacterial spectrum very similar to that of the original compounds. It is also excreted more slowly, causing blood levels to remain high for longer periods of time and making it possible to give lower doses.

D. Semisynthetic Tetracyclines Modified in Positions 6 and 7

Comparison of the activities of the natural tetracyclines and of different derivatives has shown that positions 5, 6, 7, and 9 can have different substituents without any substantial loss of activity. Obviously these are not directly involved in the formation of the bond with the biological receptor. A great deal of research led to development of two reactions that enable modifications to be made in these positions.

1. After halogenation with *N*-chlorosuccinamide in position 10, a molecule of water can be removed from position 6 to give 6-methylenetetracycline. By this procedure, methacycline(6-desoxy-6-demethyl-6-methyleneoxytetracycline) was prepared. With this as starting material, it was possible to synthesize doxycycline, or 6-desoxy-5-oxytetracycline (Fig. 5.18). However, there have been no striking biological results from this brilliant chemical work. Both methacycline and doxycycline have been introduced into therapeutic use, but again their only advantages are the prolongation of high blood levels.

2. Hydrogenolysis under specific conditions (with rhodium as catalyst) leads to loss of the hydroxyl in position 6. The activity of demethyltetra-

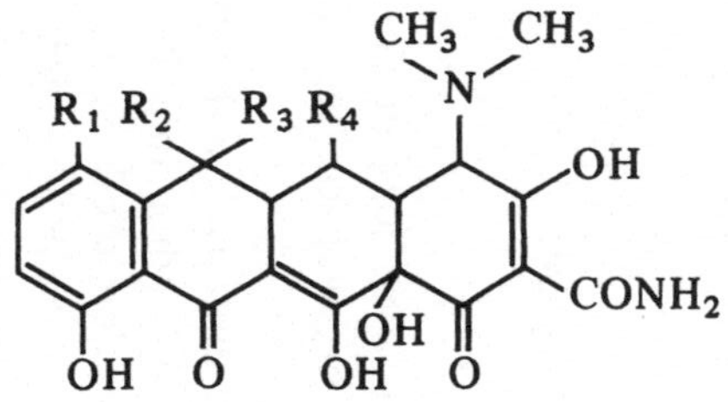

MIC (μg/ml)

R_1	R_2	R_3	R_4	Name	S. pyo- genes	S. aureus (tetracycline resistant)	S. pneumoniae	E. coli	P. mirabilis
Cl	H	OH	H	Demethylchlortetracycline	0.8	—	1	3	25
H	CH$_3$	=CH$_2$	OH	Methacycline	0.8	—	1	3	25
H	CH$_3$	H	OH	Doxycycline	0.4	—	0.04	1.6	50
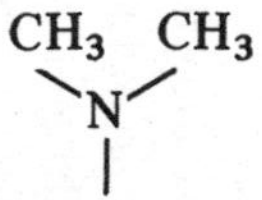 CH$_3$ CH$_3$ \N—	H	H	H	Minocycline	0.4	0.8	0.04	3.1	50

Figure 5.18. Modified tetracyclines.

cycline was essentially unchanged by this removal of the hydroxyl, but the activities of tetracycline and oxytetracycline were considerably decreased, not because the hydroxyl was eliminated but because the reaction led to the inversion of the configuration of the methyl group at position 6. In fact, doxycycline, mentioned previously, also lacks the hydroxyl at position 6, but the methyl group is in the correct configuration and the activity remains as great as that of oxytetracycline. Removal of the hydroxyl group confers great stability on tetracyclines. Therefore, 6-demethyl-6-deoxytetracycline is the ideal starting material for even drastic chemical reactions carried out to obtain derivatives with substitutions in positions 7 and 9.

Among these, the most important derivative is minocycline (Fig. 5.18) or 7-dimethylamino-6-demethyl-6-deoxytetracycline, which is active both in vitro and in vivo against strains of *S. aureus, Streptococcus,* and *E. coli* (but not *Proteus* or *Pseudomonas*) that are resistant to tetracycline. However, the hope that this compound would represent only a beginning and that other tetracyclines without cross-resistance with the original would be synthesized does not seem to have materialized. Only a very few laboratories are still working on new tetracyclines. The only significant product that has been developed in recent years is 6-thiatetracycline, at present undergoing advanced clinical trials.

E. Unresolved Problems

1. Resistance

Strains resistant to tetracyclines can be produced in the laboratory by serial culture, but the resistance develops slowly and is of the multistep type. However, in nature there are many resistant strains, principally those that carry the R factor for transferable resistance.

The low frequency of resistant mutants among the bacteria derived from the normal strains has the practical consequence that one seldom sees "conversion," i.e., the appearance of resistant bacteria, during treatment of patients who have infections with sensitive bacteria. However, the widespread diffusion, especially in hospitals, of resistant strains is the cause of an increased number of infections refractory to tetracycline treatment. It was previously stated that minocycline is active against some mutants that are resistant to other tetracyclines. Only long-term clinical practice will show whether its effectiveness is sufficiently general or whether it will be of only limited usefulness.

2. Adverse Reactions

The tetracyclines also sometimes cause hypersensitivity reactions and allergy, but with much lower frequency than the penicillins and the cephalosporins. However, gastrointestinal irritation is rather common and very high doses can be hepatotoxic. Because of their wide spectrum of action, partial absorption, and excretion in the bile, they can alter the intestinal flora to a great extent, with consequent additional infections, sometimes severe, by yeasts or by resistant *S. aureus*. Other typical secondary effects, due to the deposition of tetracycline in growing bone and teeth, is a yellow-brown discoloration of the teeth and a reversible slowing of bone growth when this antibiotic has been given to children younger than 5 years of age. Tetracyclines, especially demethylchlortetracycline, frequently cause photosensitivity with consequent erythema when the skin is exposed to sunlight. All these effects are related to the general biological properties of the tetracyclines; thus it appears highly unlikely that any active new derivative will not also have these effects.

IV. Aminoglycosides

This is a very large class of antibiotics produced by several strains of *Streptomyces, Micromonospora,* and *Bacillus.* They are quite similar in their chemical characteristics, general biological properties, and mechanism of action.

In terms of chemical structure, they are aminocyclitols (cyclohexane with hydroxyl and amino or guanidine substituents) with glycosyl substituents to one or more hydroxyl groups. Because of this structure, the molecule is basic and has a high degree of solubility in water and poor solubility in lipids. These properties make it clear why the aminoglycoside antibiotics are poorly absorbed when given orally and suggest that their transport across the bacterial membrane must take place through a specific active transport mechanism and not through simple passive diffusion.

The aminoglycoside antibiotics can be divided according to the nature of the aminocyclitol, as shown in Table 5.1.

A. Streptomycin

1. Properties

Streptomycin (Fig. 5.19) was discovered in 1944 as the result of a research program carefully planned to isolate an antibiotic that would be active against Gram-negative organisms.

In addition to having excellent activity against Gram-negative bacteria (except for some *Proteus* and some only slightly sensitive *Pseudomonas* species), streptomycin at low concentrations inhibits *Mycobacterium tuberculosis*. It was the first drug effective as a cure for tuberculosis. It is also active against staphylococci, but streptococci are rather insensitive. The spectrum of activity is reported in Chapter 2. As mentioned above, it is not

Table 5.1. Main Aminocyclitol Antibiotics

Aminocyclitol		Main antibiotics
Name	Structure	
Streptidine	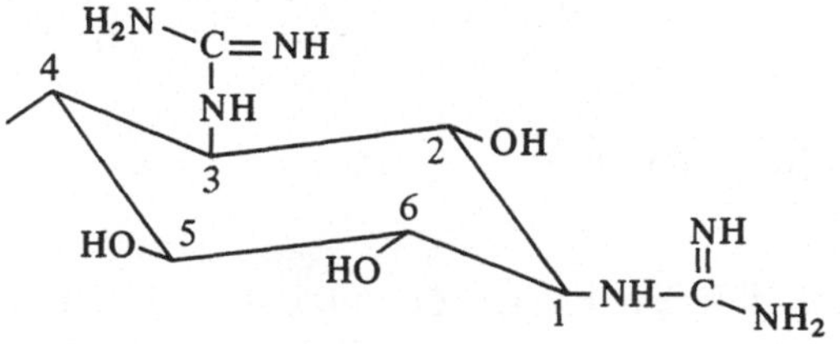	Streptomycin

$H_2N-C=NH$
NH
4
3
2 OH
6
HO 5
HO
1 $NH-C$, NH_2 (NH)

| Bluensidine | 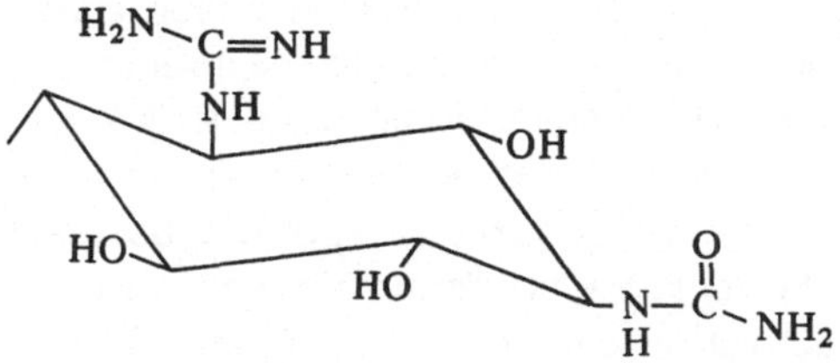 | Bluensomycin |

$H_2N-C=NH$
NH
OH
HO
HO
$N-C$, NH_2 (O, H)

2-Deoxystreptamine		Neomycin
		Paromomycin
		Lividomycin
		Kanamycin
		Gentamycin
		Sisomycin
		Ribostamycin
		Butirosin
		Tobramycin
		Seldomycin

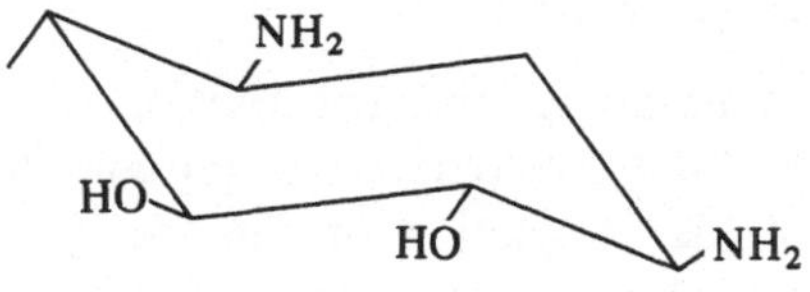

NH_2
HO
HO
NH_2

| Actinamine | 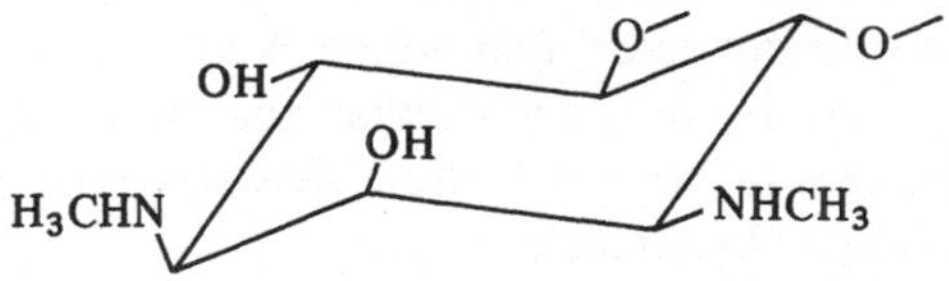 | Spectinomycin |

OH
OH
O
O
H_3CHN
$NHCH_3$

absorbed when given orally. When it is given parenterally, it is mostly excreted in the urine, which makes it suitable, as are the other amino-glycoside antibiotics, for treatment of urinary tract infections.

The negative aspects greatly limiting its use today are (1) a specific toxicity for the eighth cranial nerve that results in vestibular damage and sometimes deafness, (2) a rather high frequency of allergic reactions, and (3) a high frequency of resistant mutants and their widespread diffusion.

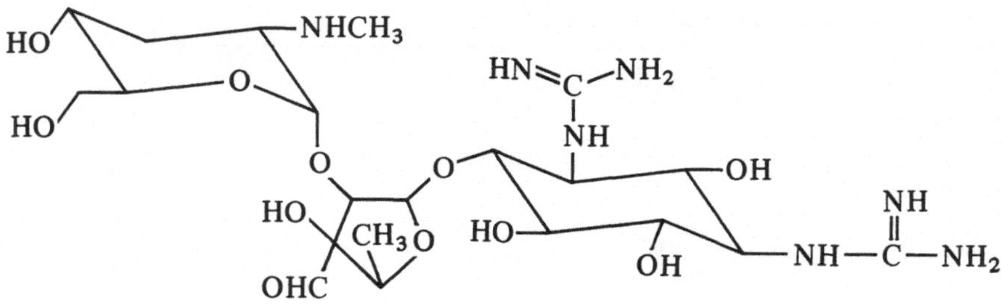

Figure 5.19. Streptomycin.

There are different classes of mutants resistant to streptomycin. The first contains those with an altered 30S ribosomal subunit, which is known to be the site of action of this antibiotic. These mutants are not resistant to other aminoglycoside antibiotics. Another class of mutants owes its resistance to decreased permeability to the antibiotic. These mutants tend to be cross-resistant with other aminoglycosides.

Both of these types of mutants are present in populations of susceptible bacteria with different frequencies and can be selected during treatment of patients. The most widely found mechanism of resistance, however, is enzymatic inactivation caused by transferable R factor, which consists of adenylation or phosphorylation of the hydroxyl group at position 3 of the methylglucosamine.

2. Structural Modifications

Elimination of the amidino groups, substitution of amine functions, and reductive amination of the aldehyde group all completely inactivate streptomycin. In contrast, demethylation of the methylamine, substitution of one aminidine group with a carbamyl (as in the antibiotic bluensomycin), and catalytic reduction of the aldehyde to a hydroxyl do not change the activity substantially. The last reaction produces dihydrostreptomycin, a compound that has also been isolated from a *Streptomyces* strain. It was introduced into therapy but later abandoned because even though it was less toxic than streptomycin for the vestibular apparatus, it more frequently caused hearing loss. It is of interest to note that reduction of the aldehyde decreases the frequency of hypersensitivity.

B. Aminoglycoside Antibiotics Containing 2-Deoxystreptamine

Many compounds are members of this group, some of which have clinical application.

2-Deoxystreptamine itself has no biological activity.

Substitution with various amino sugars in position 6 produces compounds that are still inactive. But when the substitution is in position 5 or 4, there is antimicrobial activity, even though it is weak. Good activity appears only when positions 4 and 5 or 4 and 6 are both substituted.

There are thus two series of antibiotics containing 2-deoxystreptamine that are used in therapy. In the first, the aminocyclitol is substituted in two adjacent hydroxyls (positions 4 and 5); in the second, in two nonadjacent hydroxyls (positions 4 and 6) (see Fig. 5.20). In the first series, the sugar in position 5 is almost always ribose, and when an amino sugar is attached to it, the activity increases. In both series, the activity is influenced by the presence of amine groups on the sugar in position 4. There is maximal activity when there are two amines, in positions 2' and 6', and there is less activity when an amine is present only at one of these two positions. Another structural characteristic essential for activity is the presence of the two amine groups at positions 1 and 3 on the deoxystreptamine part of the molecule. Usually any substitution of these has negative effects, but there are some specific cases, described later, in which a substitution on the amine at C-1 confers interesting properties on the molecule.

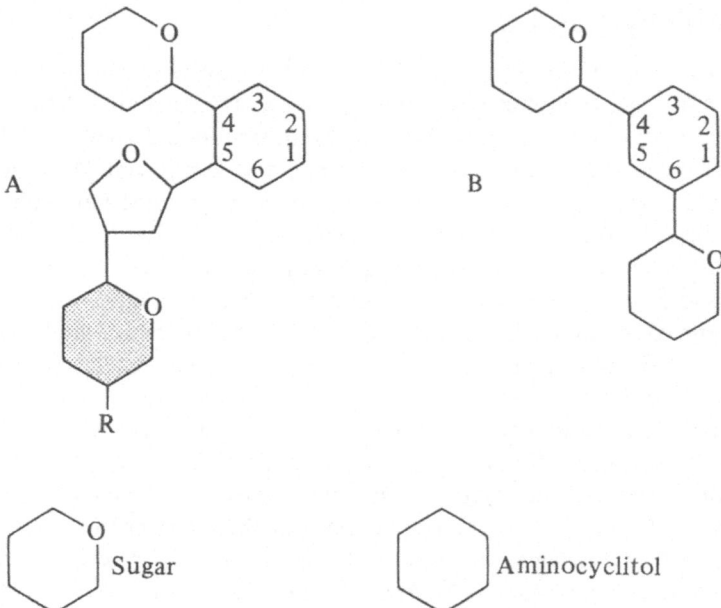

Figure 5.20. Schematic representation of the two classes of antibiotics containing 2-deoxystreptamine. (**A**) 4,5-Substituted. The *shaded* sugar is missing in the ribostamycins. An additional sugar is present at position R in lividomycins. (**B**) 4,6-Substituted. In both classes the antibiotics differ in the amino, hydroxy, and methyl substituents on the sugars.

1. Derivatives of Deoxystreptamine Substituted in Positions 4 and 5

The best known products in this category are the neomycins, a mixture of antibiotics, the major one of which is neomycin B, isolated in 1949 from *Streptomyces fradiae* (see Fig. 5.21). They have a spectrum of activity similar to that of streptomycin, including the modest degree of activity against *Streptococcus* and *Pseudomonas*. Neomycin is used only topically because it is ototoxic and nephrotoxic. Paromomycin (aminosidine) resembles neomycin in its chemical structure and biological properties. It is used as an intestinal disinfectant because it is not absorbed into the general circulation when taken orally. It is used for treatment of intestinal amebiasis as well as for bacterial infections.

Two antibiotics for systemic use, ribostamycin and lividomycin A, have been developed recently in Japan. The first is less active than the other deoxystreptamine derivatives (because the ribose is unsubstituted), but it is also less toxic. The second has a good spectrum of activity, even against *Pseudomonas,* and is less toxic than neomycin.

2. Derivatives of Deoxystreptamine Substituted in Positions 4 and 6

This group of antibiotics (Fig. 5.22) was discovered relatively recently and has become even more important because of the activity of the natural products and of some recently developed semisynthetic derivatives. The first product in this group to be used in therapy was kanamycin A, which was isolated in 1957 from *Streptomyces kanamyceticus,* together with the B and C derivatives. Its spectrum of activity is very similar to that of streptomycin, but it is less toxic to the ear, though equally toxic to the kidneys. It is used for Gram-negative infections and in Japan it is also used for treatment of tuberculosis. Kanamycin B, or becanamycin, which has biological properties similar to those of kanamycin A, was recently introduced into clinical use.

In 1963 the gentamycin complex, which contains many components, was isolated from *Micromonospora purpurea.* The term gentamycin is commonly used for the mixture used for therapy and it consists of gentamycins C_1, C_{1a}, and C_2, which differ only in their degrees of methylation of carbon 6 of the sugar in position 4. Gentamycin is very active against *S. aureus* and Gram-negative bacteria. Its excellent activity against *Pseudomonas* and several *Proteus* species is of particular importance, and it is widely used for these purposes. However, it is used only for severe infections, because of vestibular and nephrotoxicity.

More recent products introduced into clinical use are tobramycin and sisomycin. The first is produced by *Streptomyces tenebrarius* and has a structure similar to that of kanamycin (it is 3′-deoxykanamycin B) but a spectrum of activity like that of gentamycin. It is more active than gentamy-cin against *Pseudomonas* and somewhat less active against the other Gram-negative bacteria. Sisòmycin is produced by *Micromonospora inyoensis* and has a structure resembling gentamycin and very similar biological properties, except that it is more effective against *Pseudomonas*.

3. Relationships Between Structure and Enzymatic Inactivation and Semisynthetic Derivatives

To understand the relationship between the structures of the derivatives of 2-deoxystreptamine and their activity against certain bacterial strains, it is necessary to examine the mechanisms of resistance against these anti-biotics in some strains or species.

Unlike streptomycin, no cases of resistance due to alterations of the target, the ribosomal proteins, have been found, probably because these antibiotics can bind to more than one ribosomal site. Resistance due to inability to penetrate the bacterial cell are also infrequently seen, except in certain *Pseudomonas* strains. But quite frequently strains are isolated that produce enzymes able to inactivate the antibiotic. An example is shown in Fig. 5.23, which schematically presents the modifications produced in kanamycin B by bacterial enzymes.

Obviously, not all of these are of practical importance. Acetylation of the nitrogen in position 6' occurs rarely and is insignificant, adenylation of the oxygen in position 4' and phosphorylation of the oxygen in position 2'' are carried out only by *S. aureus,* a species for which aminoglycosides should be prescribed only in exceptional cases. On the other hand, phos-phorylation of the oxygen in position 3' is quite common, especially by *Pseudomonas* strains, and two other important reactions are the acetyla-tion of the amine in position 3 of deoxystreptamine and nucleotide attach-ment to the oxygen in position 2'. One can therefore understand the activity of tobramycin (3'-deoxykanamycin B) against *Pseudomonas,* since this antibiotic does not possess the hydroxyl group in position 3' and there-fore cannot be inactivated by phosphorylation by the *Pseudomonas* en-zymes. In the same way, gentamycin cannot be phosphorylated in position 3' or have the nucleotide attached in position 4' because neither of these two carbon atoms are hydroxylated.

This information served as the basis for rational planning of a program for synthesizing chemically modified derivatives. Among these, dibekacin (3',4'-deoxykanamycin B) and amikacin (*N*-hydroxy-γ-aminobutyrylkana-mycin A) have been adopted for clinical use (Fig. 5.24). Dibekacin is of course insensitive to the enzymes that attack the hydroxyls in positions 3' and 4' and therefore has activity comparable to that of gentamycin. The synthesis of amikacin was suggested by the observation that an antibiotic produced by *B. circulans,* butirosin B, differing from ribostamycin only in having an α-hydroxy-γ-aminobutyric acid substituent on the amine in po-

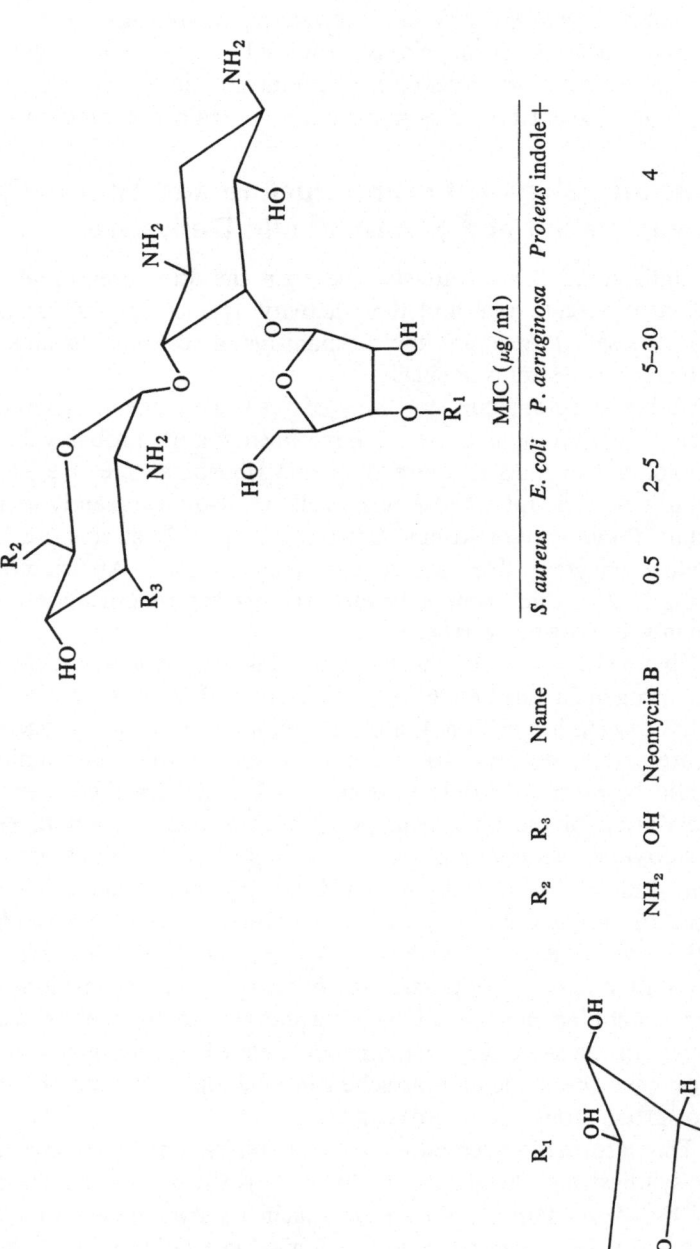

R_2	R_3	Name	S. aureus	E. coli	P. aeruginosa	Proteus indole+
				MIC (μg/ml)		
NH_2	OH	Neomycin B	0.5	2–5	5–30	4

R₁	R₂	R₃	Name	MIC (μg/ml)			
				S. aureus	E. coli	P. aeruginosa	Proteus indole+
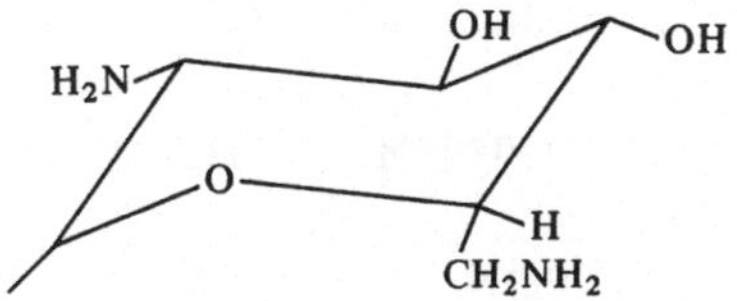	OH	OH	Paromomycin I	1	8	>100	4
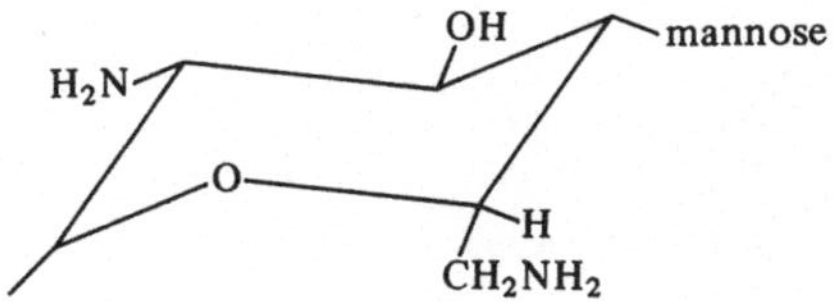	OH	H	Lividomycin A	0.8	6	6	3
H	NH₂	OH	Ribostamycin	12.5	3–6	>100	3–6

Figure 5.21. Neomycin B, paromomycin I (aminosidine), lividomycin A, and ribostamycin.

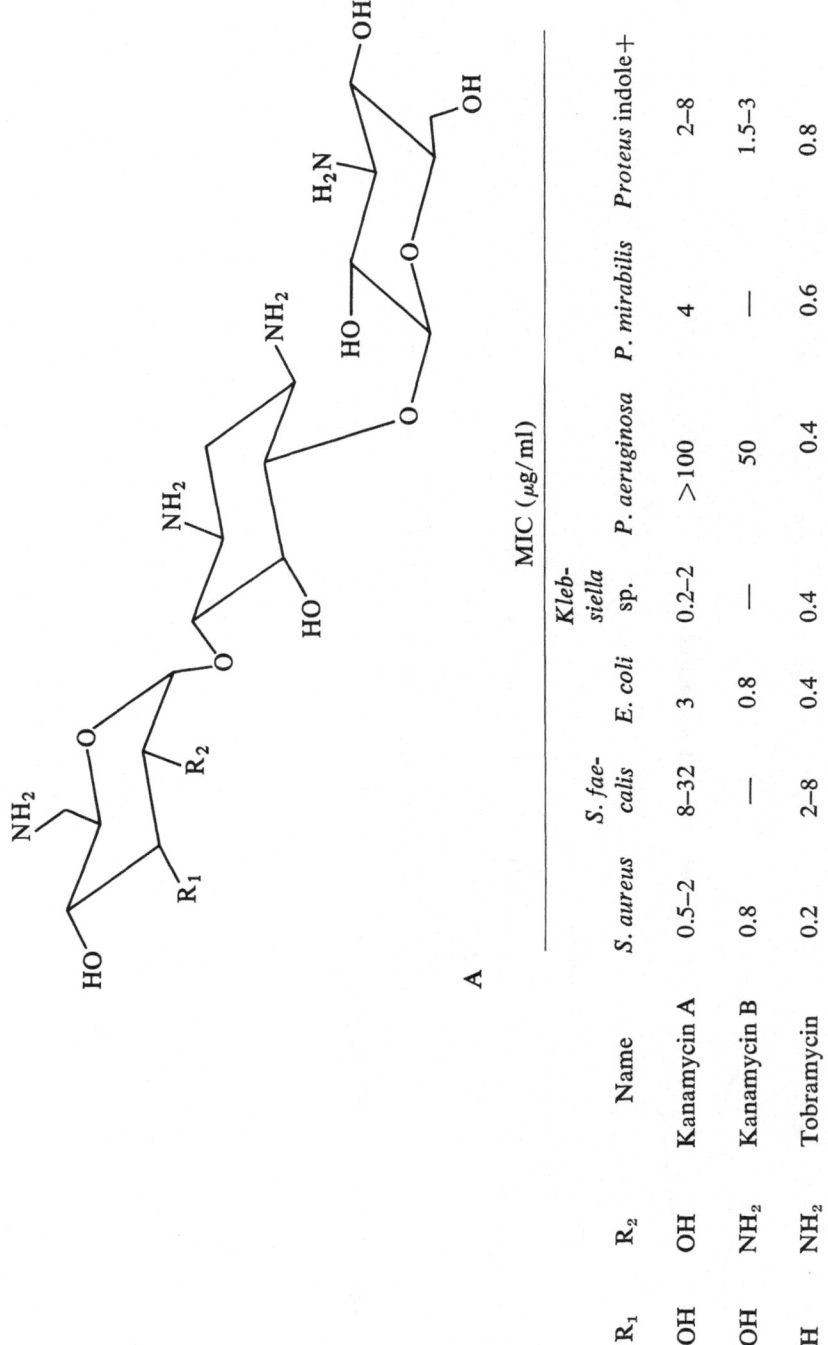

A

MIC (μg/ml)

Name	R₁	R₂	S. aureus	S. fae-calis	E. coli	Kleb-siella sp.	P. aeruginosa	P. mirabilis	Proteus indole+
Kanamycin A	OH	OH	0.5–2	8–32	3	0.2–2	>100	4	2–8
Kanamycin B	OH	NH₂	0.8	—	0.8	—	50	—	1.5–3
Tobramycin	H	NH₂	0.2	2–8	0.4	0.4	0.4	0.6	0.8

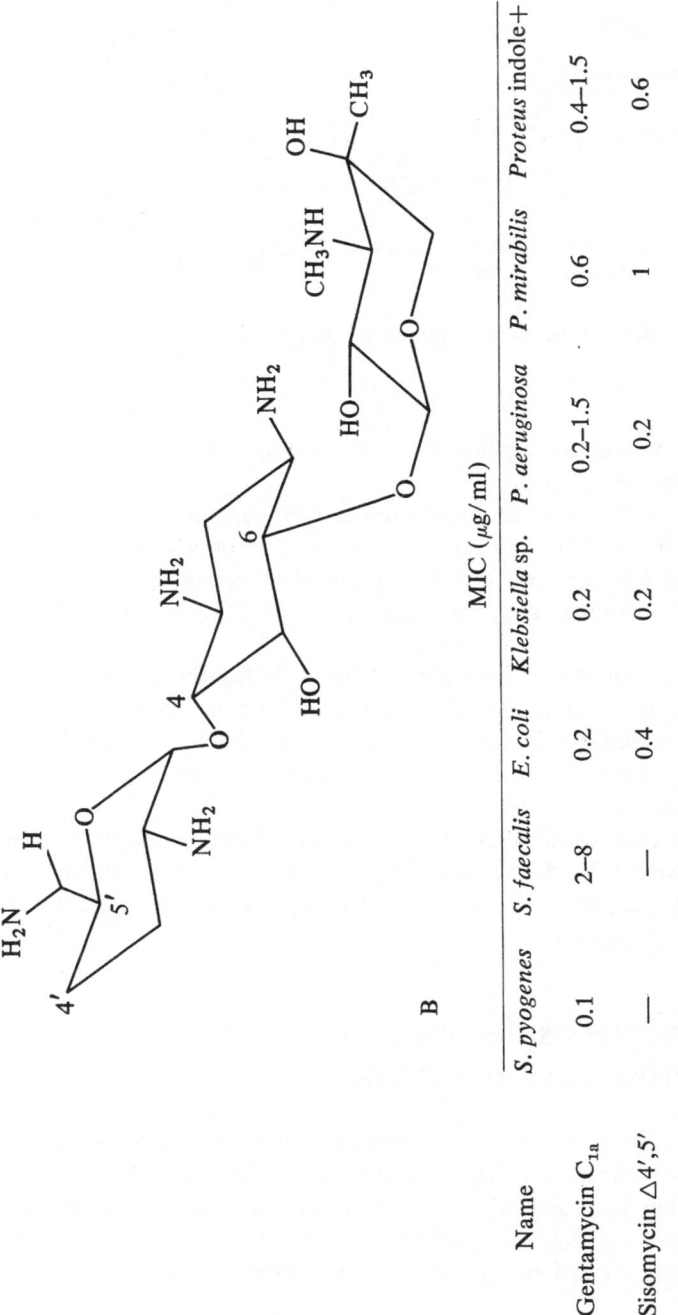

Name	S. pyogenes	S. faecalis	E. coli	Klebsiella sp.	P. aeruginosa	P. mirabilis	Proteus indole+
Gentamycin C_{1a}	0.1	2–8	0.2	0.2	0.2–1.5	0.6	0.4–1.5
Sisomycin $\triangle 4',5'$	—	—	0.4	0.2	0.2	1	0.6

MIC (μg/ml)

Figure 5.22. (**A**) Kanamycin A, kanamycin B, and tobramycin. (**B**) Gentamycin C_{1a} and sisomycin.

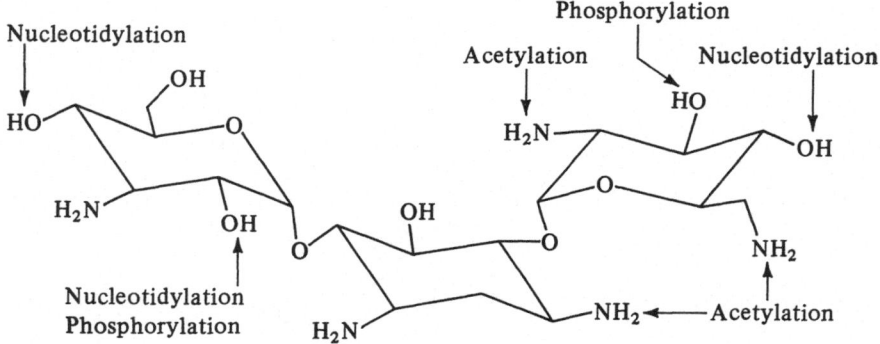

Figure 5.23. Representation of points of inactivation of kanamycin B by bacterial enzyme.

sition C-1, was active against *Pseudomonas* and other inactivating strains, unlike ribostamycin itself.

When the different aminoglycosides were acylated with α-hydroxy-γ-aminobutyric acid, the molecules were found to be insensitive to the phosphorylating enzymes for the 3′ oxygen, the acetylating enzymes for the position 3 nitrogen, and the nucleotide attaching enzymes for the 2″ oxygens.

Amikacin, the derivative obtained by acylating the nitrogen on C-1, of kanamycin A, is the aminoglycoside with the broadest spectrum of activity. It is used especially in cases of infection by bacteria resistant to other derivatives, since its activity against sensitive strains is less than that of gentamycin.

Another recently developed semisynthetic derivative is netilmicin, 1-*N*-ethylsisomycin (Fig. 5.24). It has a somewhat broader spectrum of activity than does gentamycin. Its principal advantage is the lower toxicity it shows in laboratory animals.

C. Aminoglycoside Antibiotics Containing Different Aminocyclitols

In addition to those already described, only one other aminoglycoside antibiotic is currently being used clinically. This is spectinomycin (Fig. 5.25), which is produced by *Streptomyces spectabilis* and contains the aminocyclitol actinamine. It differs from the other derivatives in that it is bacteriostatic instead of bactericidal. It is used for the treatment of gonorrhea.

There are other aminoglycoside antibiotics that are used in agriculture. Probably the most important of these is kasugamycin, used in rice culti-

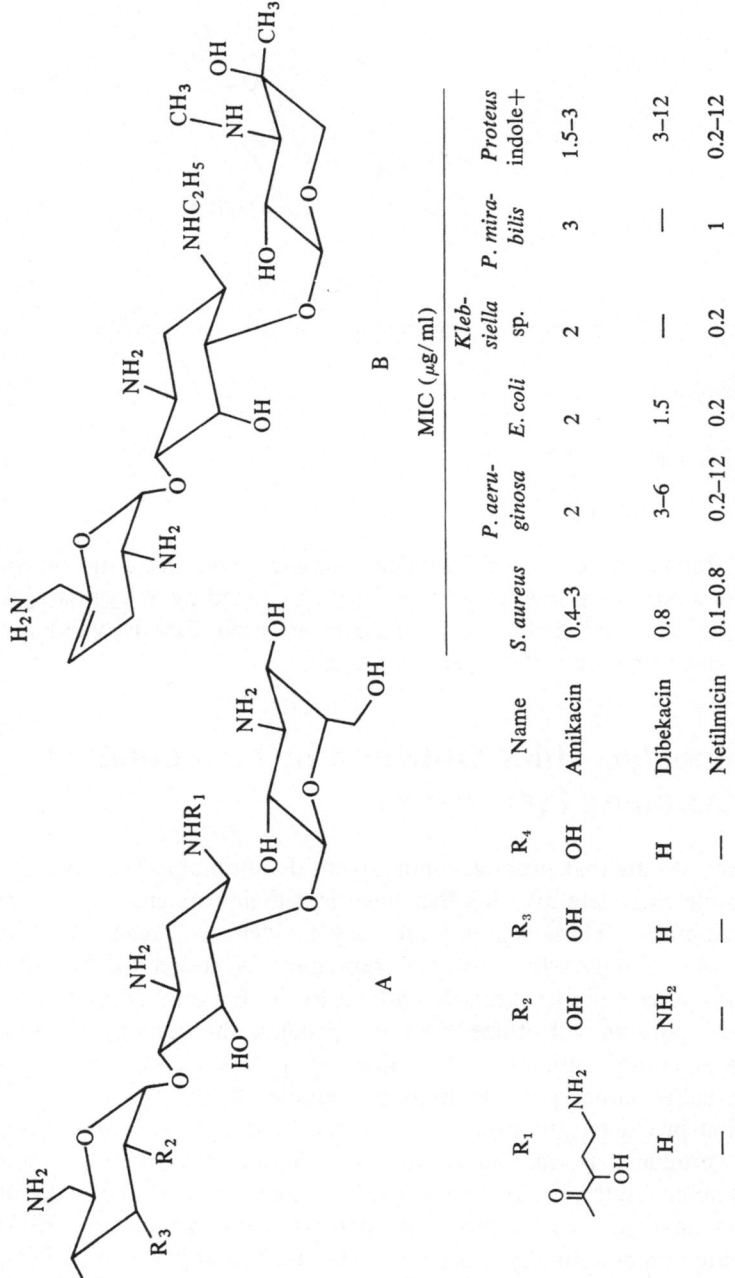

Figure 5.24. Semisynthetic aminoglycosides. (**A**) Amikacin and dibekacin. (**B**) Netilmicin.

Name	R₁	R₂	R₃	R₄	S. aureus	P. aeruginosa	E. coli	Klebsiella sp.	P. mirabilis	Proteus indole+
					MIC (µg/ml)					
Amikacin	$\begin{array}{c} O \\ \parallel \\ \diagdown NH_2 \\ OH \end{array}$	OH	OH	OH	0.4–3	2	2	2	3	1.5–3
Dibekacin	H	NH₂	H	H	0.8	3–6	1.5	—	—	3–12
Netilmicin	—	—	—	—	0.1–0.8	0.2–12	0.2	0.2	1	0.2–12

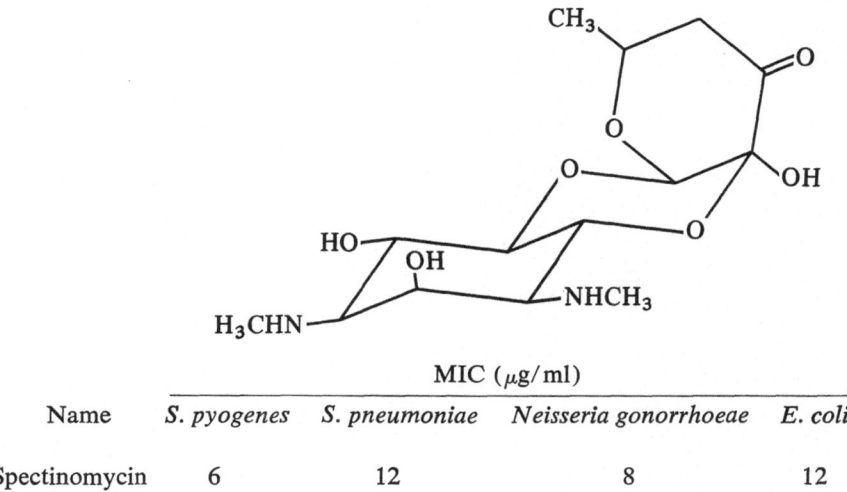

MIC (μg/ml)

Name	S. pyogenes	S. pneumoniae	Neisseria gonorrhoeae	E. coli
Spectinomycin	6	12	8	12

Figure 5.25. Spectinomycin.

vation in Japan for control of *Piricularia oryzae* infections. For control of ascarid infections in animals, hygromycin B, produced by *S. hygroscopicus,* and destomycin, produced by *S. rimofaciens,* are used. Finally, validomycin is used against some fungal diseases of plants.

D. Aminoglycosides Obtained by Fermentation of Synthetic Precursors

By treating strains that produce aminoglycoside antibiotics with mutagens, it is possible to isolate mutants that have lost their capacity to synthesize deoxystreptamine. These mutants are easy to identify, because they produce antibiotics only when 2-deoxystreptamine is added to the culture. If, instead, other aminocyclitols are added to the fermentation, sometimes the microorganisms will utilize them and produce the corresponding analogs of the natural antibiotics. This method for production of new antibiotics is called either mutasynthesis or mutational biosynthesis.

The first products thus obtained were the hybrimycins (2-hydroxyneomycins), produced by mutant strains of *S. fradiae* grow in the presence of streptamine. Later, other microorganisms were used to synthesize new antibiotics analogous to streptomycin, paromomycin, kanamycin, ribostamycin, butirosin and, finally, sisomycin. The method appears to be fruitful. The products obtained, many of which have microbiological activity, have helped to clarify the structure–activity relationships and some of them may prove to be of interest for therapeutic use.

One disadvantage of this method is that, at least to date, the yields are too low for industrial production.

E. Unresolved Problems

The spectra of activity for these antibiotics have been discussed when describing the individual compounds. It should be noted that in general this class of antibiotics lacks activity against anaerobic bacteria or aerobic bacteria growing under anaerobic conditions.

We have already mentioned the problem of resistance and the attempts to resolve it by planned synthesis of new products. There is no doubt that the new derivatives are an important therapeutic advance, and it is possible that current research efforts, which are very active in this field, will yield further results.

Because of the general characteristics of the class, it is not very likely that derivatives will be prepared that can be absorbed by oral administration.

The most important question in this field is whether or not it will prove possible to biosynthesize or synthesize antibiotics of this family that are less toxic than the ones we have today. We cannot definitely answer this question until we understand at the biochemical level the reasons for the toxicity toward the eighth cranial nerve and the kidneys. Generally speaking, analysis of the effects of various substituents on the toxicity and comparisons of the activity and the toxicity of tens of products seem to indicate that activity and toxicity change in parallel.

Should the animal studies indicating that netilmicin is less toxic be confirmed in man, there will be a great enhancement of the impetus to search for new nontoxic aminoglycosides.

V. Macrolides

Antibiotics with structures containing large aliphatic lactones are called macrolides. They are divided into two classes with quite different biological properties and structures: The antibacterial macrolides and the antifungal macrolides, also known as the polyene macrolides.

A. Antibacterial Macrolides

This group is characterized chemically by the comparatively smaller size of the lactone rings, from 12 to 16 atoms (Fig. 5.26) and biologically by the mechanism of action, the specific inhibition of protein synthesis in the bacteria due to formation of a complex with the ribosomal 50S subunit.

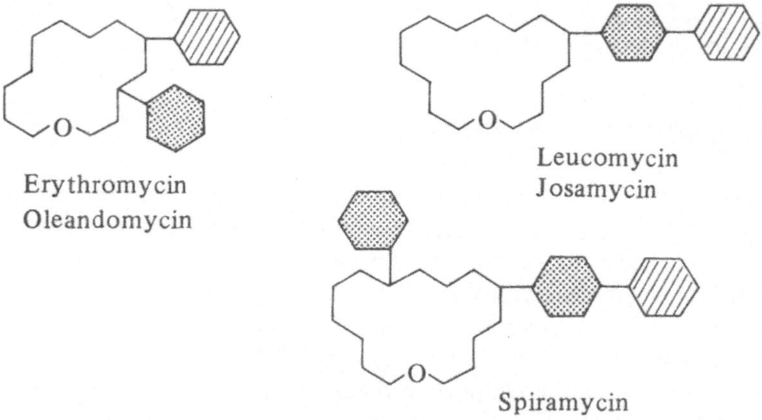

Figure 5.26. Schematic representation of the structure of clinically used macrolides: *unshaded figures,* lactonic ring; *striped figures,* amino sugars; *dotted figures,* sugars. All other substituents and glycosidic bonds have been omitted.

The lactone ring has methyl and hydroxyl substituents in various positions. At least one glycoside substituent is always present, usually in position 5, and frequently a second one in position 3. The macrolides can be either neutral or basic, depending on whether the glycosides are ordinary sugars or amino sugars (this latter is much more common).

The spectrum of antibacterial activity resembles that of penicillin G, with good activity against Gram-positive and some Gram-negative cocci. With some exceptions, discussed later, they have cross-resistance among themselves and with lincomycin.

Some of them have been introduced into therapy. These are erythromycin, carbomycin (no longer used), oleandomycin, spiramycin, kitasamycin (or leucomycin) and, recently, josamycin. Tilosin, which is effective against mycoplasma, is widely used in veterinary medicine. The antibacterial activities of the principal macrolides are presented in Table 5.2.

Table 5.2. Activity of Macrolide Antibiotics Against Selected Bacteria

	MIC (μg/ml)				
Test organisms	Oleando-mycin	Erythro-mycin	Spira-mycin	Kitasa-mycin	Josa-mycin
S. aureus	0.5	0.2	2	0.4	0.4
S. pyogenes	0.2	0.02	0.2	0.4	0.2
S. pneumoniae	0.2	0.1	0.2	0.04–0.2	0.8
Mycoplasma pneumoniae	0.6	0.1	0.6	0.3	0.03

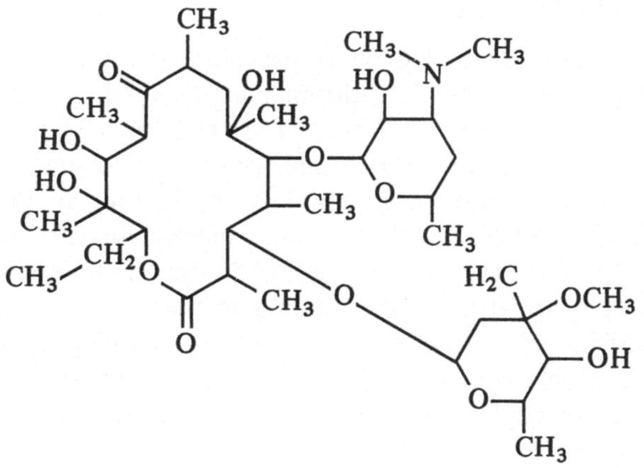

Figure 5.27. Erythromycin A.

1. Erythromycin

This is the most widely used of the macrolide antibiotics. It was isolated in 1952 from fermentations of *Streptomyces erythreus* as a complex of three very similar substances, erythromycins A, B, and C, of which the most active and major component was erythromycin A (Fig. 5.27). Its structure is typical of the macrolides, with a 14-atom ring, no double bonds, methyl groups on the even-numbered carbon atoms, and a carbonyl in position 9. Position 5 is substituted with desosamine and position 3 with cladinose.

It is quite active against Gram-positives and is used for streptococcal and pneumococcal infections. It is less active against staphylococci, which are often resistant. Its activity against Gram-negatives, such as *E. coli,* is demonstrable in vitro but is inadequate for therapeutic purposes. It is usually given orally, but one of the problems with erythromycin is erratic absorption, because it is not very stable in an acid environment (the first macrolide used in therapy, carbomycin, had to be abandoned because its absorption was too unreliable). On the other hand, the preparations for intramuscular administration, e.g., erythromycin ethylsuccinate, are not satisfactory because the injection is painful. To increase the blood levels (which after oral administration of erythromycin base are below 0.5 μg/ml), frequent use is made of the laurylsulfate of erythromycin propionyl ester, called erythromycin estolate. This derivative gives blood levels at least three times higher, but since the propionic ester is inactive, only the fraction hydrolyzed in vivo, which is difficult to measure quantitatively, can be taken into consideration. In addition, with the estolate, and only with it, some cases of hepatotoxicity have occurred, almost certainly due to hypersensitivity.

a. Structure–Activity Relationships

Several semisynthetic derivatives were prepared and their activities compared with those of the natural erythromycins. It was found that (1) both of the glycoside substituents were needed for activity; (2) modification of the desosamine, especially of the amine group, led to inactive products, probably because this group is involved in the binding to the ribosomes (esters of the adjacent hydroxyl at 2' were inactive in vitro but active in vivo because they are easily hydrolyzed); (3) the cladinose does not appear to be directly involved in the binding to the ribosomes, since it can be partly modified without inactivation; and (4) modification of the functions in the positions from 9 to 12 of the lactone ring, reduction or substitution of the carbonyl group, elimination of the hydroxyl group at 12, acetylation in 11 give products with some activity.

b. Resistance

Two types of mutants resistant to erythromycin have been described:

1. Those with resistance associated with a mutation in one ribosomal protein. These are found in several bacterial species. The mutants are resistant not only to erythromycin but also to other macrolides and to lincomycin.
2. Those with resistance due to an alteration (methylation) of the ribosomal RNA. This change is induced by the presence of small quantities of erythromycin or of oleandomycin but not of the other macrolides. When it is present, the bacteria are resistant to all the macrolides and to lincomycin, but when the inducer is removed from the culture, the mutant again becomes susceptible. This type of resistance has been found only in staphylococci.

2. Oleandomycin

Oleandomycin was isolated in 1954 from *Streptomyces antibioticus*. It is a macrolide similar to erythromycin in structure and in biological properties. It is given orally in an acetylated form, such as triacetyloleandomycin, which is absorbed better than the natural product. It has limited use, as it offers no advantages over erythromycin and appears to be on the whole less active.

3. Spiramycin, Leucomycin (Kitasamycin), Josamycin

These are macrolides with 16-atom lactone rings, and, like the other members of this class, have an acetaldehyde group in position 6 and one pair of conjugated double bonds in the ring. They have a disaccharide in position 5 (Fig. 5.26). Spiramycin (Fig. 5.28) differs from the others in

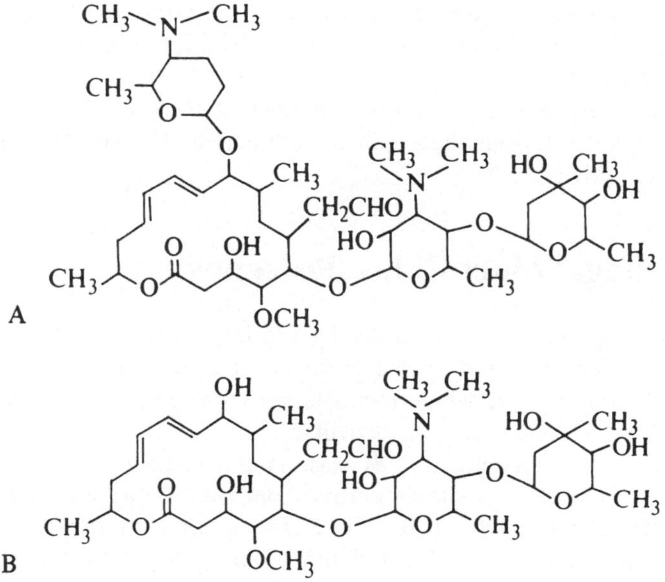

Figure 5.28. (A) Spiramycin A. (B) Leucomycin A_1 (kitasamycin).

having an amino sugar on position 9 as well. Leucomycin is a mixture of several products, the major ones being A_1 (Fig. 5.28), isolated in 1953, and spiramycin, isolated in 1955. Josamycin is very similar to and possibly identical with leucomycin A_3, but it was isolated independently from a different microorganism in 1964.

These products have antibacterial spectra similar to that of erythromycin. They are more resistant in acidic environments but slightly less active in the classic experimental septicemia when given orally.

They have been introduced into therapy in some countries because of their activity against staphylococci with acquired resistance to erythromycin. In all other strains, there is cross-resistance among them and with other macrolides.

Some chemical modification studies have shown that active products can be obtained after hydrogenation of the two double bonds or hydrolysis of the second sugar. Reduction of the aldehyde group decreases the activity, but it can be converted to the hydrazone derivative without any loss of the original activity.

4. Unresolved Problems

The side effects of the macrolides are not very frequently seen and are usually slight. The principal problems with this class of antibiotics seem to be unsatisfactory absorption and the spread of resistant strains. The problem of absorption may be resolved by preparation of semisynthetic derivatives or isolation of new natural macrolides, for which an active search is continuing.

It is not likely that striking improvement of the resistance will be obtained. It is unlikely that there will be any active attempts to increase the spectrum of activity.

B. Antifungal Macrolides (Polyenes)

These are characterized structurally by the large size of the lactone ring, from 26 to 38 atoms, and by the presence of a series of conjugated double bonds, from 3 to 7, which characterize them as polyenes, in general, and as trienes, tetraenes, etc., in particular.

A general characteristic of this class is the presence in the ring of a series of hydroxyls, in positions across from the double bonds (see, e.g., the structure of amphotericin B (Fig. 5.29), so that the molecule has two distinct zones, one hydrophilic and the other hydrophobic, and is therefore a surfactant. Other common substituents are a carboxyl group and a sugar, mycosamine.

Polyenes are typically active against fungi, sometimes against some protozoa, and only exceptionally against bacteria. This is a consequence of their mechanism of action, which is alteration of the function and the integrity of the cell membrane by complexing with the sterols that are components of eukaryotic membranes.

The antimicrobial activity increases with the number of conjugated double bonds. There are not yet sufficient data to establish whether or not the toxicity, which is quite high, increases in parallel with the activity.

It has been shown that there is no correspondence between hemolytic activity and antifungal activity. Several polyenes are used for treatment of infections of the skin and the mucosa and also of the intestine, since after oral administration they are not absorbed into the circulation. Among these are pymaricin, tricomycin, candicidin, and nystatin. The last is by far the most widely used. It is a tetraene, but with respect to both the size of the ring and other structural details it significantly resembles the heptaenes.

1. Amphotericin B

This is a heptaene (Fig. 5.29) with a 38-carbon ring produced by *Streptomyces nodosus*. It is very active against several pathogenic fungi and is used intravenously for cure of severe systemic infections. Treatment with amphotericin B is used only in life-threatening conditions, because of the toxicity of the antibiotic, which manifests itself principally in the kidneys, often causing irreversible damage.

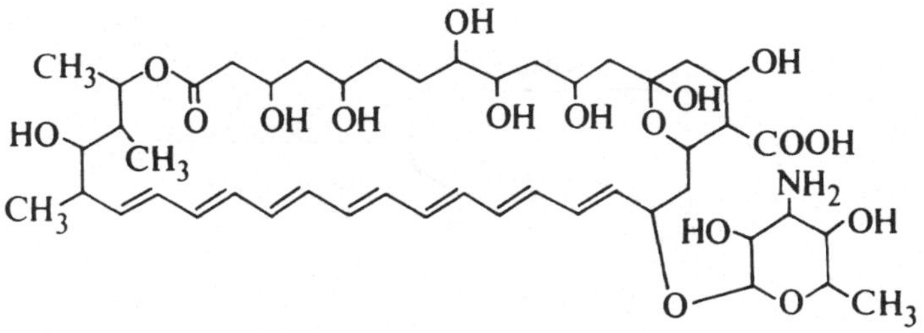

Figure 5.29. Amphotericin B.

Oral and local administration, obviously, do not carry the same risk. It is known that oral administration leads to a decrease in absorption of cholesterol, which complexes with the antibiotic in the intestine.

2. Chemical Modifications

It is usually thought that the structural features making the polyenes active are the same as those that make it toxic, and therefore not much has been done to improve the therapeutic index by making chemical changes. However, it has recently been reported that the methyl esters of amphotericin and of other macrolides retain their antifungal activity and, given as the soluble hydrochlorides, should be less toxic.

3. Resistance

Although it is possible to obtain in the laboratory strains resistant to the polyene macrolides (with cross-resistance among them), from the clinical point of view resistance not yet presented any problems. The resistant strains have alterations in the structure of the cell membrane.

VI. Ansamycins

The ansamycins are the most recent family of antibiotics to acquire clinical importance. They are characterized by a cyclic structure that in some ways resembles that of the macrolides. The cyclic portion consists of an aromatic group and an aliphatic chain and the ring is closed by an amide group, making them lactams and not lactones.

They can be divided on the basis of the aromatic group into benzene ansamycins and naphthalene ansamycins (Fig. 5.30). The benzene ansamycins, which include geldanamycin and maytansin, have been isolated more recently. They differ from the naphthalene ansamycins in that their

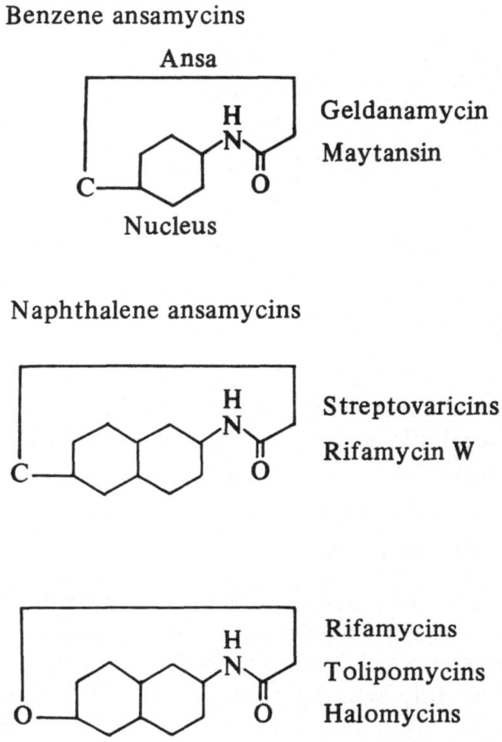

Benzene ansamycins

Naphthalene ansamycins

Figure 5.30. Schematic representation of ansamycin structure.

activity is not selectively antibacterial; they are toxic. At present they are being studied as potential antitumor agents.

The naphthalene ansamycins are a rather homogeneous chemical class and are characterized biologically by their mechanism of action, which is specific inhibition of RNA synthesis in bacteria by formation of a complex with RNA polymerase.

Depending on the producing microorganism and some structural details, they are divided into the streptovaricins (in which the aliphatic chain is directly attached to the aromatic nucleus by a carbon–carbon linkage) and the rifamycins, tolipomycins, and halomycins (in which an ether bond interrupts the continuity between the aliphatic chain and the aromatic group.)

Among the ansamycins, the most important group is the rifamycins (Fig. 5.31). These are a remarkable example of how chemical modifications can improve the therapeutic properties of natural antibiotics. One semisynthetic rifamycin, rifampin (Fig. 5.32), is the most active antibiotic available today for treatment of tuberculosis.

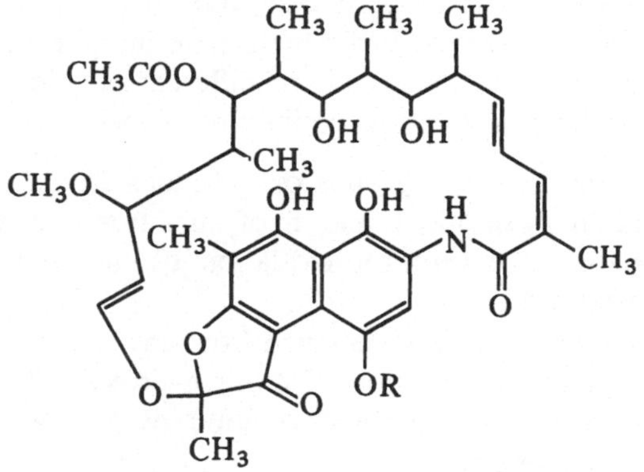

Figure 5.31. Rifamycin SV (R=H) and rifamycin B (R=CH$_2$COOH).

A. Natural and Semisynthetic Rifamycins

The first naturally occurring rifamycins were isolated in 1958, as a complex of five active substances, from fermentations of *Streptomyces mediterranei* (now called *Nocardia mediterranea*). Under specific culture conditions it is possible to obtain a single component from the fermentation, rifamycin B (Fig. 5.32), which is moderately active against Gram-positive bacteria and mycobacteria and which was tested for possible clinical use because of its low toxicity. It was found that in dilute solution its activity increased with time, suggesting that it could be transformed into more active derivatives. This was the first step in the preparation of semisynthetic rifamycins. The first of these to be introduced into therapy was rifamycin SV (Fig. 5.32), obtained by removing the glycol chain in position 4 of rifamycin B by oxidative cyclization, hydrolysis, and reduction of the quinone. Rifamycin SV can today be considered a natural rifamycin, since programmed research, guided by biosynthetic considerations, led to isolation of a mutant of *N. mediterranea* that produces it directly. It has very effective and rapid bactericidal action against Gram-positive bacteria and mycobacteria. It is less active against Gram-negatives. It is well tolerated and is given parenterally, especially for infections of the bile tract, where it reaches high enough concentrations to be effective even against the Gram-negatives. The limitations of rifamycin SV (which have dictated objectives for continuing research into semisynthetic rifamycins) are essentially as follows:

1. ineffectiveness when given orally;
2. weak activity against Gram-negative bacteria;
3. rapid elimination in the bile, with consequent instability of tissue levels.

This last characteristic is of particular importance in tuberculosis, for which rifamycin SV was found to be ineffective in spite of its optimal in vitro activity. The modifications made in different positions on the molecule enabled the following structure–activity relationships to be established.

1. Changes in the hydroxyl groups on positions 21 and 23 of the aliphatic chain and in positions 1 and 8 of the aromatic nucleus always give inactive products (except for oxidation of the hydroxyl group at position 1 to a carbonyl).

2. Inactive products were also obtained from any structural modification that changed the spatial orientation of those hydroxyls. Therefore, it appears that these participate in the formation of the bonds with RNA polymerase (see Chapter 3).

3. Minor modifications of the chain, such as hydrogenation of the conjugated double bonds, deacetylation, etc., do not modify the activity greatly.

4. Substitutions on positions 3 or 4 of the aromatic nucleus do not alter the intrinsic activity of the molecule, but they can alter the physiochemical properties, and consequently such biological properties as absorption, distribution, and penetration into bacteria, especially the Gram-negative bacteria. If the substituent includes a carboxyl group, the antibacterial activity is invariably low, whereas basic functions increase the activity against Gram-negative bacteria.

5. Activity after oral administration is often found in derivatives with substitutions in position 3 or with cyclic substituents containing a nitrogen function in positions 3 and 4. It is occasionally found in those substituted only in position 4.

Practical results of this research include rifamide (the diethylamide of rifamycin B), which has a better therapeutic index than rifamycin SV even though it has essentially the same limitations, and rifampin, called rifampicin outside the U.S. (Table 5.3).

Table 5.3. Activity of Rifamycins Against Selected Bacteria

Test organism	MIC (μg/ml)		
	Rifamycin SV	Rifamide	Rifampin
S. aureus	0.005	0.01	0.002
S. faecalis	0.05	0.1	0.01
S. pneumoniae	0.025	0.02	0.01
K. pneumoniae	25	20	5
E. coli	50	10	1
Pseudomonas aeruginosa	50	50	10
Proteus vulgaris	25	20	5
Mycobacterium tuberculosis	0.05	0.2	0.05

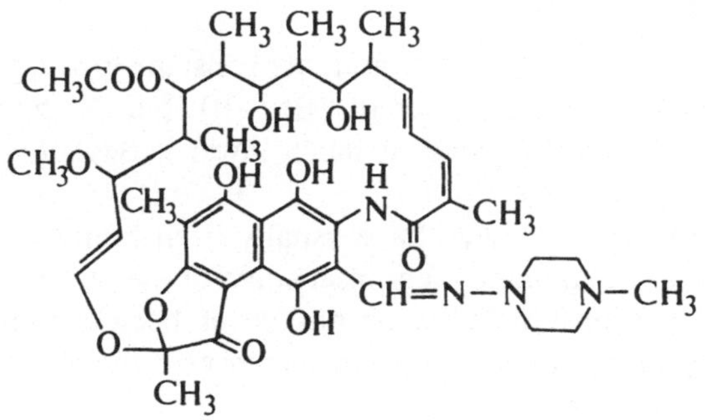

Figure 5.32. Rifampin.

Rifampin (Fig. 5.32) has been found to possess a high degree of the desired characteristics. It has a broad spectrum of activity, is absorbed when given orally and, most important, because of its better distribution in the body, is also very effective for treatment of tuberculosis. It is used throughout the world for treatment of this disease and in several countries for other infections as well. Recently it has been suggested for treatment of leprosy.

Among the many semisynthetic rifamycins, some substituted in position 3 with long lipophilic chains have been found to have some degree of inhibitory activity against an enzyme typical of oncogenic viruses, the so-called reverse transcriptase (an enzyme used by RNA viruses to synthesize complementary DNA strains). Many other derivatives with analogous characteristics have been synthesized and enormous effort has been expended in study of their biological properties (still in progress), without any practical outcome. There has also been no practical consequence of the observation that at high concentrations rifampin inhibits the reproduction of certain DNA viruses such as vaccinia.

B. Unresolved Problems

1. Side Effects and Hypersensitivity

Rifampin rarely causes side effects except for a harmless increase in blood bilirubin levels. However, some patients treated over periods of many months, especially when treatment has been intermittent, develop hypersensitivity, sometimes of very serious degrees.

Studies are underway to determine whether or not the other rifamycins have the same mechanism of action and to determine whether antigenic activity is due to rifampin itself or one of its metabolites.

2. Resistance

There is a rather high frequency of mutants resistant to rifamycins in some bacterial populations, of the order of 10^{-7}–10^{-8} in *Staphylococcus* and *E. coli*. Fortunately, the frequency is much lower in *Mycobacterium tuberculosis* (10^{-10}).

The resistant mutants studied have usually been found to have changes in RNA polymerase that make them insensitive to the antibiotic. Resistance associated with a transferable factor has never been demonstrated. This is probably why there has not been an increase in the spread of resistant strains after several years of use.

The rifamycins, of course, show cross-resistance within the group. Some lipophilic derivatives have been demonstrated to have some degree of activity against staphylococci resistant to rifampin, but not great enough to be of therapeutic interest.

It is unlikely that the problem of resistance can be overcome by making rifamycin derivatives. It is more likely to be circumvented by administering it in suitable combinations with other antibiotics.

VII. Peptide Antibiotics

As indicated by the name, these are antibiotics comprised of amino acid chains connected through peptide bonds. They are of importance historically because almost all of the initial antibiotics isolated and studied for therapeutic use belonged to this class.

They are a highly heterogeneous group chemically and differ from proteins because of their smaller molecular weight (on the order of thousands of daltons) and in the following frequently found characteristics:

1. They contain D-amino acids.
2. They contain some uncommon amino acids (*N*-methylaminoacids, β-aminoacids, etc.).
3. They often are cyclic molecules.
4. They contain heterocyclic rings, frequently thiazoles.

The group is even more heterogeneous in its biological activities: antibacterial activities, the mechanisms of action, and toxicity. They can be classified into subgroups, since some structural characteristics are associated with certain mechanisms of action, but there are not yet sufficient data to do this validly.

The structure–activity relationships have been studied in detail for only two families of polypeptide antibiotics These are the penicillins–cephalosporins and the actinomycins. (The penicillins and cephalosporins have already been discussed. Chemically they can be classified among the polypeptide antibiotics.)

The actinomycins have limited use as antitumor agents. They inhibit the synthesis of both RNA and DNA by binding to DNA, and the structural requirements for formation of that complex have been studied a great deal in the hope of preparing more selective derivatives.

A. Systemically Used Polypeptide Antibiotics

Most peptide antibiotics are too toxic for systemic use. The following exceptions are noteworthy:

1. Polymyxin B and Colistin

These are interesting examples of antibiotics active only against Gramnegative bacteria. They are produced by various strains of bacilli and are very similar in their chemical structures (Fig. 5.33) and biological properties, such as antibacterial activity and toxicity. Polymyxin B is generally used as the sulfate for intramuscular or intravenous injections, whereas the preferred form for colistin is its methane sulfonate derivative, which is less toxic but correspondingly less active. Both are used only for treating severe infections by organisms such as *Pseudomonas* that are resistant to other less toxic antibiotics. The emergence of resistant strains is rare.

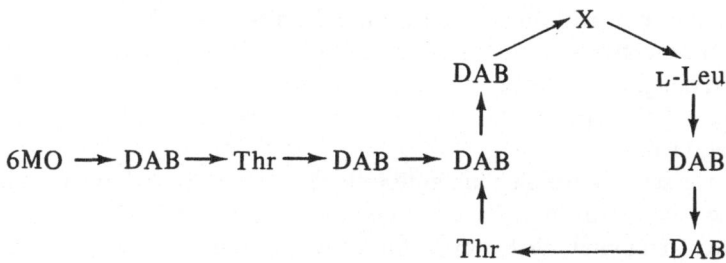

6-MO = 6-Methyloctanoic acid

DAB   = 2,4 Diaminobutyric acid

Thr   = Threonine

Leu   = Leucine

⟶ = Peptide bond (CO-NH)

Figure 5.33. Structure of polymyxin B (X=D-phenylalanine) and of colistin (X=D-leucine).

2. Capreomycin

This antibiotic was isolated in 1960 from *Streptomyces capreolus*. It is active only against mycobacteria. Its MIC against *M. tuberculosis* is somewhat high, about 10 μg · ml^{-1}, but since high concentrations are easily attained in blood the product can be used in the treatment of tuberculosis patients when the infecting strain is resistant to other less toxic agents.

3. Bleomycin

Bleomycin is a product of *Streptomyces verticillus*. Nine natural products and more than 100 semisynthetic derivatives are known. They are basic glycopeptides that inhibit DNA synthesis and are used in the treatment of carcinomas, Hodgkin's disease, and lymphomas.

B. Topically Used Peptide Antibiotics

These include some of the earliest antibiotics studied for therapeutic use. Tyrothricin, isolated in 1939, is a mixture of two products, gramicidin (about 20%) and tyrocidin (about 80%). Gramicidin is very active against Gram-positive bacteria. Tyrothricin is less active but also has some activity against Gram-negative bacteria. Both are hemolytic agents, which makes it impossible to give them systemically. Topical tyrothricin is used for infections of the eye, nose, and throat and for ulcers and wounds.

Bacitracin was isolated in 1943 and is active against Gram-positive bacteria. It interferes with the synthesis of cell walls. It can be given intramuscularly in exceptional cases but is normally considered too toxic (producing kidney damage) for systemic use. It is of some use in treatment of open infections of the skin or the cornea. It is very widely used as an additive to animal feed.

Uses for several polypeptide antibiotics have been found outside clinical medicine, such as nisin, a food preservative, and micamycin and thiostrepton, used in animal husbandry. An active search is underway for new polypeptides to be used in these areas. And one cannot exclude categorically that new antibiotics of this class will prove to be useful in systemic treatment of man.

VIII. Miscellaneous Antibiotics

Under this heading are grouped some antibiotics used clinically that do not fit into any of the previously described classes either because of their chemical or their biological properties. We also describe two classes of antibiotics that are not discussed in any detail because they are not antimicrobial, but solely antitumor, agents. These are the nucleosides and the anthracyclines.

1. Nucleosides

These are analogs of the natural nucleoside, from which they differ either in the base portion (purine or pyrimidine) or in the glycoside portion. They are not used as antibacterial agents because they are antagonists of the natural nucleosides in many enzymatic raections, and therefore are active not only in bacterial cells but also in those of the higher animals. Among the many that have been described, we mention only puromycin, which is used in the laboratory as an inhibitor of protein synthesis, an activity based on its structural similarity to the terminal portion of amino-acyl tRNA.

2. Anthracyclines

These are glycosidic derivatives of a four-ring linear structure that resembles the tetracyclines. They differ markedly from tetracyclines in their mechanism of action. In fact, they form complexes with DNA and interfere in that way with both its transcription and its replication. The most important anthracyclines are daunomycin and adriamycin (Fig. 5.34), which are used as antitumor agents.

3. Chloramphenicol

This was isolated in 1947 from a strain of *Streptomyces venezuelae*. Chloramphenicol was the first therapeutic agent to be effective against rickettsiae and related agents, as well as against the major Gram-negative bacteria and some Gram-positives. It is well absorbed after oral administration. It is the drug of choice for treatment of typhoid and salmonellosis and among the most effective for curing infections with *Hemophilus influenzae, Klebsiella pneumoniae,* etc. The antibacterial spectrum of chloramphenicol is presented in Chapter 2, Table 2.3.

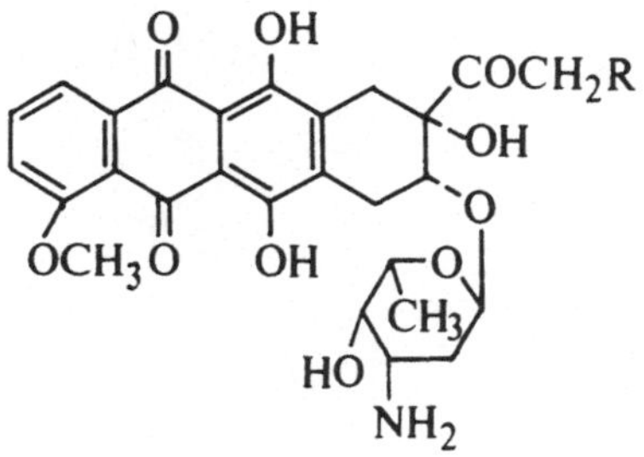

Figure 5.34. Daunomycin (R=H) and adriamycin (R=OH).

Its therapeutic use is limited by the risk of bone marrow aplasia, a hypersensitive reaction that is nearly always fatal in the rare cases (about one in 40,000 treated) in which it occurs. The occurrence of aplasia is not correlated with the dose but appears after long treatment and can be prevented by constant monitoring of the hematological condition of the patient. Less serious side effects include anemia, which is reversible and dose dependent, some gastrointestinal disturbances, and, because of the broad spectrum of the antibiotic's action, secondary infections resulting from destruction of the normal bacterial flora. Very high doses are dangerous to small infants because they may not excrete it efficiently.

The structural formula for chloramphenicol is shown in Fig. 5.35. Because there are two asymmetric centers, four stereoisomers are possible. Among these, only the D-threoisomer is active. Chloramphenicol is the only antibiotic produced industrially by synthesis instead of by fermentation. From the numerous analogs that have been synthesized, the following structure–activity relationships could be inferred.

1. The dichloroacetic acid residue can be replaced by other acids without substantial loss of activity as long as they are strongly electronegative and do not cause more steric hindrance.

2. The propanediol structure is essential for activity as indicated by the fact that only one stereoisomer has activity. Esters of the primary hydroxyl retain activity, probably to the degree to which they can be hydrolyzed in vivo. The palmitate is used because it has no taste.

3. In the aromatic portion, the nitro group is not indispensable and can be replaced by other polar functions, such as a methylthio or methylsulfonic group. The thiamphenicol derivative (Fig. 5.35) obtained by this last replacement is used in therapy with considerable success.

Resistance to chloramphenicol is usually due to inactivating enzymes that acetylate one or both of the hydroxyls. These enzymes are inducible

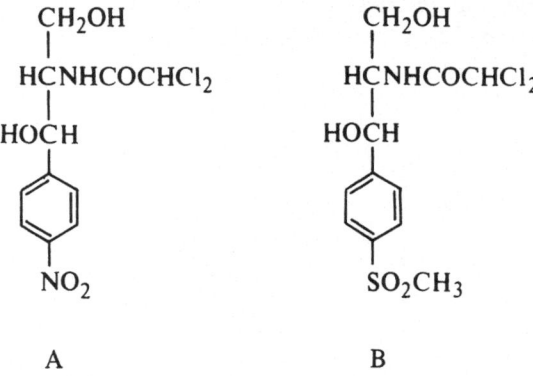

Figure 5.35. (A) Chloramphenicol. (B) Thiamphenicol.

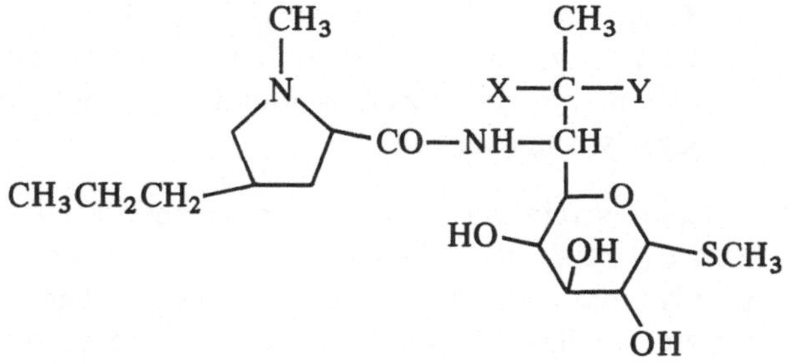

			MIC (μg/ml)			
X	Y	Name	*S. aureus*	*S. pyogenes*	*S. pneumoniae*	*E. coli*
OH	H	Lincomycin	0.5	0.05	0.5	>100
H	Cl	Clindamycin	0.1	0.02	0.05	>100

Figure 5.36. Lincomycin and clindamycin.

in *S. aureus* and constitutive in *E. coli*. In *E. coli,* the enzyme is coded for by the R factor, which frequently carries the determinants for resistance to other antibiotics.

4. Lincomycin and Clindamycin

Lincomycin, isolated in 1962 from the fermentation broth of *S. lincolnesis,* is active against Gram-positive bacteria. Chemically (Fig. 5.36), it can be considered an amino sugar acylated with a proline derivative. Its biological activity resembles that of the antibacterial macrolides (it inhibits protein synthesis by complexing with the ribosomal 50S subunit), and it has some degree of cross-resistance with them. It has been suggested that in spite of the apparent structural differences, the three-dimensional model of lincomycin has points of similarity with erythromycin. Three of the methyl groups of erythromycin (that on the ethyl, that on carbon 2, and that in position 5 of the cladinose) could correspond spatially with three methyl groups of lincomycin (the S-methyl, the methyl on group 8, and the methyl on the propyl group). In addition, the hydroxyl in position 12 of the macrolides could coincide with the hydroxyl in position 2 of lincomycin. Because of the simplicity of the molecule, several analogs have been prepared semisynthetically, with the following results:

1. The acyl portion can be modified and still have active products. The activity increases when the size of the alkyl group on position 4 of the proline is increased.
2. When the methyl on the sulfur is replaced with other alkyl groups, one obtains products that are more active in vitro but not in vivo.

3. Replacement of the hydroxyl in position 7 with a chlorine causes a marked increase in activity if the replacement takes place with inversion of the configuration. But if the configuration is maintained, the activity is unchanged.

Clindamycin was prepared by means of this modification (Fig. 5.36) and it is being used clinically because of its greater activity against Gram-positive and anaerobic bacteria. It has also been shown to have interesting activity against *Plasmodium berghei*. The increased antibacterial activity of clindamycin seems to be due to its better penetration into the bacterial cell. The antibiotics have equal activities in cell-free systems. The lack of activity of these antibiotics against Gram-negative organisms appears to be due to a lesser sensitivity of the ribosomes of Gram-negative bacteria.

The limitations of these antibiotics are more or less the same as those for erythromycin. However, intramuscular administration of lincomycin is better tolerated. They often cause side effects, for the most part intestinal disturbances, that sometimes can be serious (hemorrhagic colitis).

5. Novobiocin and Coumermycin

Novobiocin (Fig. 5.37), discovered in 1956, is active against Gram-positive bacteria and some strains of *Proteus*. It is active when given orally, but its use is decreasing because of several side effects and its modest effectiveness.

The coumermycins are chemically and biologically related to novobiocin. They are a family of antibiotics first described in 1965 that have been subjected to a considerable program of semisynthesis. The most effective natural compound, coumermycin A, has also been studied in the clinic, but it proved to be difficult to administer and not very effective. One of the semisynthetic derivatives was more active, but cannot be given to man because of hepatotoxicity.

6. Fusidic Acid

This antibiotic is active against Gram-positive organisms. It was isolated in 1962 from cultures of a fungus, *Fusidium coccineum*. It has a steroid type of structure (Fig. 5.38), as do some other antibiotics produced by fungi (e.g., cephalosporin P and helvolic acid). Several derivatives have

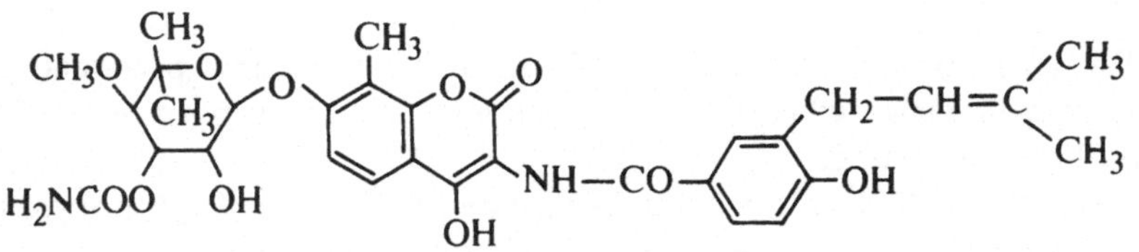

Figure 5.37. Novobiocin.

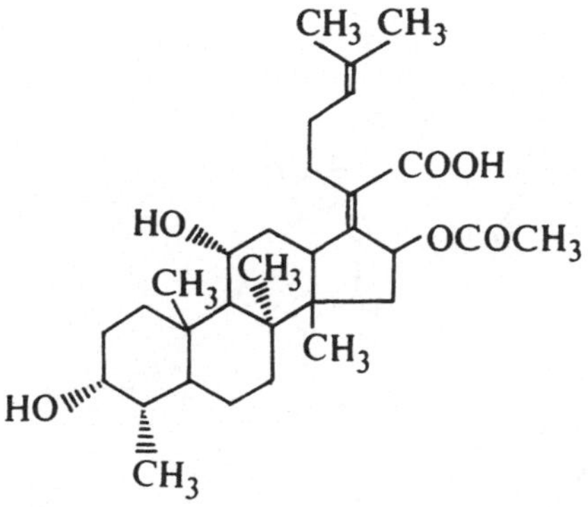

Figure 5.38. Fusidic acid.

been prepared but without success. It is absorbed when given orally and is frequently given in combination with penicillin.

7. Griseofulvin

This metabolic product of *Penicillium griseofulvum* was isolated and described in 1939 (Fig. 5.39). For several years it was thought to be inactive. In 1947 it was identified with the "curling factor," an antifungal activity produced by other penicillia.

Only in 1958 was its clinical effectiveness in skin mycoses noted, and it was introduced into therapy. It is active in vitro against the dermatophytes (*Microsporium, Epidermophyton, Trichophyton*) but not against other pathogenic fungi. Griseofulvin acts systemically by concentrating in the deep cutaneous layers and in the keratin cells. It is usually well tolerated. The most common side effect is transient headache.

Numerous semisynthetic derivatives of griseofulvin have been made, but none have activity equal to or better than that of the original product.

8. Vancomycin

Vancomycin was isolated from the fermentation broth of *Streptomyces orientalis* in 1956. It is a complex glycopeptide molecule, containing glucose and the amino sugar vancosamine and the amino acids N-methyleu-

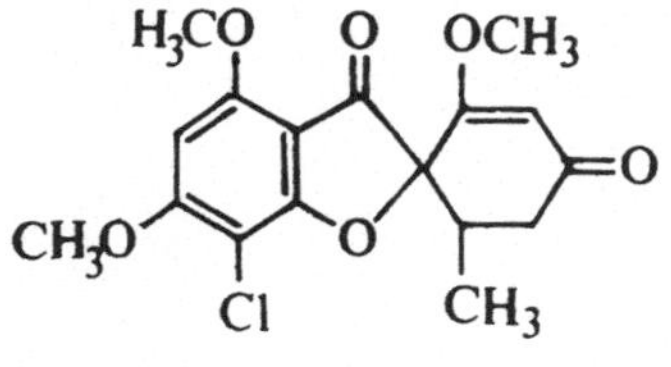

Figure 5.39. Griseofulvin.

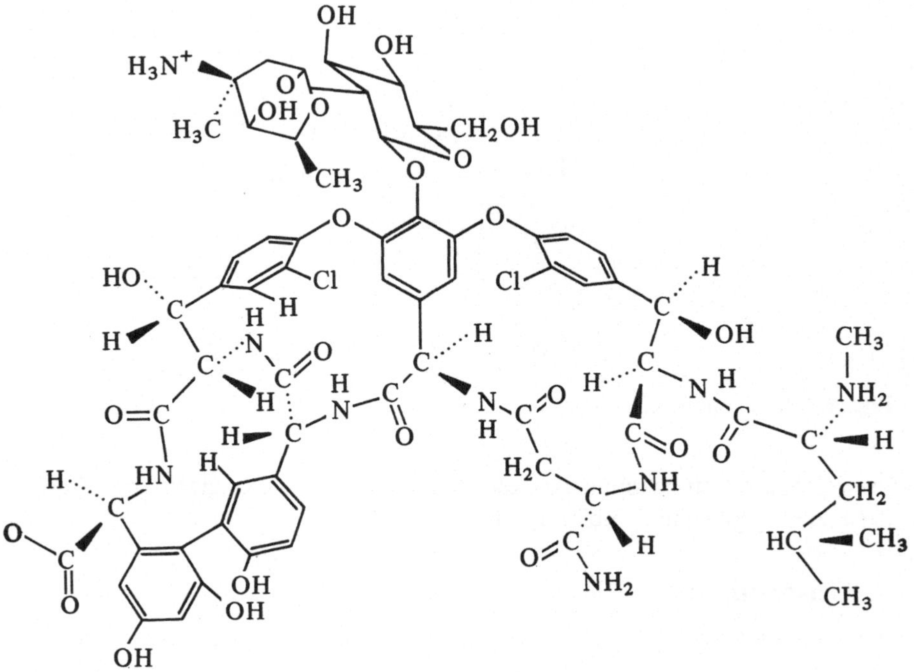

Figure 5.40. Vancomycin.

cine, aspartic acid, phenylglycine, and chloro-β-hydroxytyrosine (Fig. 5.40). It is a bactericidal drug against growing bacteria. No resistant mutants have ever been isolated. It is given intravenously to treat life-threatening staphylococcal infections refractory to other antibiotics. It is also used orally to treat staphylococcal enterocolitis. More recently, it has been shown to cure and prevent *Clostridium difficile*–induced colitis associated with use of other antibiotics, especially lincomycin and clindamycin.

The major drawbacks of vancomycin are the lack of oral and intramuscular activity (other than for treatment of staphylococcal enterocolitis), and phlebitis at the site of injection.

Other members of this family of antibiotics are: ristocetins, ristomycins, mannopeptins, avoparcin, actinsiden, teichomycins. Chemically, they differ among themselves in the type and number of sugar residues; the types of aminoacids; the number of Cl atoms, and the net charge of the molecule.

Chapter 6

Biosynthesis of Antibiotics

I. Primary and Secondary Metabolites

A. Definition

In terms of biogenesis, antibiotics are considered secondary metabolites. Secondary metabolites are natural products of low molecular weight with the following characteristics:

1. They are synthesized by only some microbial species.
2. They have no obvious function in the growth of the cell and are often produced after the culture has ceased to grow.
3. Cells able to make these molecules easily lose through mutation the capacity to synthesize them.
4. They often are made in families of similar products.

Primary metabolites are normal products of cell metabolism such as amino acids, nucleotides, coenzymes, etc. that are essential for cell growth.

B. Relationship Between Primary and Secondary Metabolism

The study of biosynthesis of antibiotics consists of identifying the series of enzymatic reactions by which one or more primary metabolites (or intermediates in the biosynthesis of these) are converted to the antibiotic molecule. It should be kept in mind that production of a secondary metabolite, especially in large quantity, implies a considerable change in the primary metabolism of the cell, which must synthesize the starting material and provide energy, for example, in the form of ATP and reduced coenzymes.

Therefore, it is not surprising that when one compares strains that produce antibiotics with strains that do not there are considerable differences in the levels of enzymes not directly involved in the antibiotic's synthesis.

C. Principal Biosynthetic Pathways

The enzyme reactions that lead to synthesis of antibiotics do not differ fundamentally from those that lead to synthesis of primary metabolites. They can be considered only variations of these, with of course some exceptions (for example, some antibiotics contain a nitro group, a function that is never found in primary metabolites and that is derived through a special pathway of amine oxidation).

We can classify the biosyntheses of antibiotics in three major categories:

1. Antibiotics derived from a single primary metabolite. The biosynthetic pathway consists of a series of reactions that modify the starting material in the same way as in synthesis of amino acids or nucleotides.

2. Antibiotics derived from two or three different primary metabolites, which are modified and condensed to give a complex molecule. There are some analogous cases in primary metabolism in synthesis of certain coenzymes, such as folic acid or coenzyme A.

3. Antibiotics derived from polymerization of several similar metabolites to give a basal structure that can then be further modified by additional enzyme reactions.

Four classes of antibiotics are derived from polymerization processes:

1. polypeptide antibiotics (derived by condensation of amino acids);
2. antibiotics built up of acetate-propionate units (by polymerization mechanisms similar to those that give rise to fatty acids);
3. terpenoid antibiotics (derived from acetate units through isoprenoid synthesis);
4. aminoglycoside antibiotics (made through condensation reactions similar to those that make polysaccharides).

Note that these processes have some similarities with the polymerization processes that provide some of the components of the membrane and of the cell wall.

It should be emphasized that the basal structure obtained by polymerization is usually modified by further reactions, even by addition of molecules made through other biosynthetic pathways. Glycosidic antibiotics made by condensation of one or more sugars onto a molecule biosynthesized by pathway (2) are particularly common.

D. Synthesis of "Families" of Antibiotics

Frequently microbial strains produce several chemically and biologically related antibiotics that constitute a "family of antibiotics." The capacity to produce "families" of compounds is not unique to antibiotic biosynthesis but is a general characteristic of secondary metabolism attributable to the fairly large size of the intermediate molecules. The biosynthesis of a family of compounds is the consequence of the following series of metabolic events:

1. Biosynthesis of a "key" metabolite through one of the pathways described in the previous section.
2. Modification of the key metabolite by rather common reactions, for example, oxidation of a methyl group to an alcohol and of the latter to a carboxyl group; reduction of a double bond; dehydration; methylation; esterification; etc.
3. The same metabolite can be a substrate for two or more such reactions, generating two or more different products which, in turn, may be enzymatically transformed, giving rise to a "metabolic tree."
4. The same metabolite can be obtained through two or more different pathways in which only the order of the enzymatic reactions is changed, thus giving rise to a "metabolic net."

The rather peculiar concepts of metabolic tree and metabolic net may be clarified by the following examples: the biogenesis of the rifamycin family (tree) and that of the erythromycin family (net).

The first metabolite in the biogenesis of the rifamycin family is protorifamycin I (Fig. 6.1), which may be considered the key metabolite.

Through a series of reactions whose order is unknown protorifamycin I is converted into rifamycin W and rifamycin S, thus completing the monopathway part of the synthesis (the "trunk" of the tree).

Rifamycin S is the starting point for the "branching off" of several alternative pathways: condensation with a C-2 moiety gives rise to rifamycin O and rifamycins L and B. The latter, through oxidation of the ansa chain, generates rifamycin Y. The oxidative elimination of C-1 of rifamycin S gives rise to rifamycin G, while a series of unknown reactions convert S to the so-called rifamycin complex (rifamycins A, C, D, and E). Oxidation of the methyl group at C-30 gives rise to rifamycin R.

The key metabolite of the erythromycin family is erythronolide B (Er. B), which is converted into erythromycin A (the most complex metabolite) by means of four reactions (Fig. 6.2):

1. Glycosidation in position 3, through condensation with mycarose (Myc) (Reaction I).
2. Conversion of mycarose to cladinose (Clad.) by methylation (Reaction II).

3. Conversion of erythronolide B into erythronolide A (Er. A) through hydroxylation in position 12 (Reaction III).
4. Condensation with desosamine (Des) in position 5 (Reaction IV).

Since the sequence of these four reactions may vary, different metabolic pathways are possible and taken together constitute the "metabolic net" described in Fig. 6.2. It must be noted that there also exist pathways that are combinations of tree and net.

II. Methods of Study

A. Tracer Techniques

As mentioned before, the study of antibiotic biosynthesis consists of identifying the reactions that convert a product of primary metabolism into

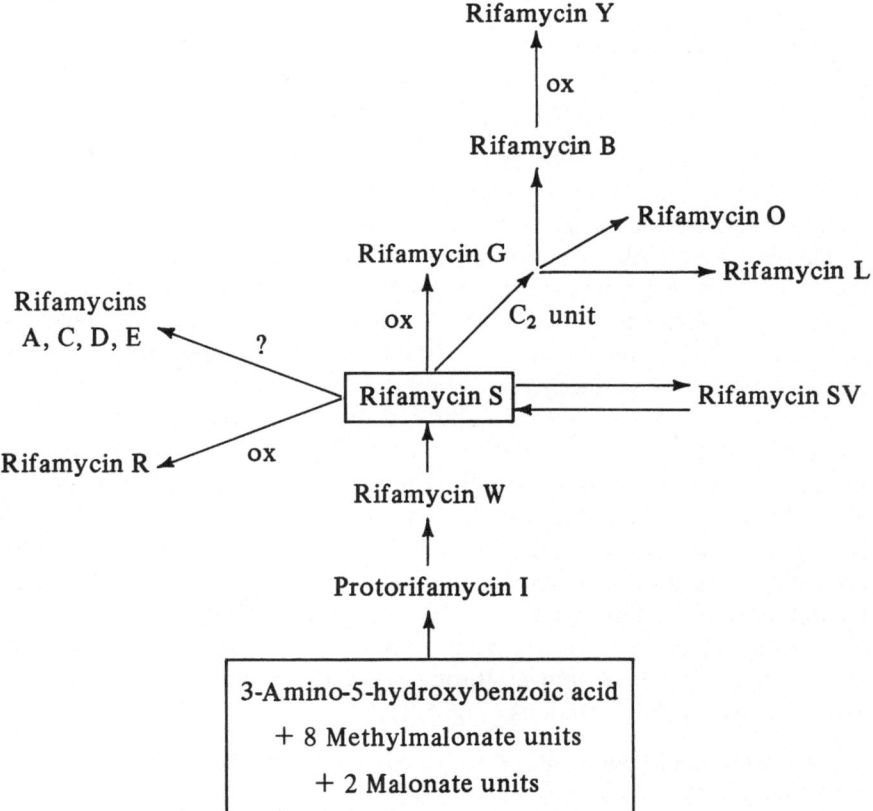

Figure 6.1. Example of metabolic tree. Rifamycin biosynthesis. (See text for explanation. Structures are shown in Figs. 6.17 and 6.23.)

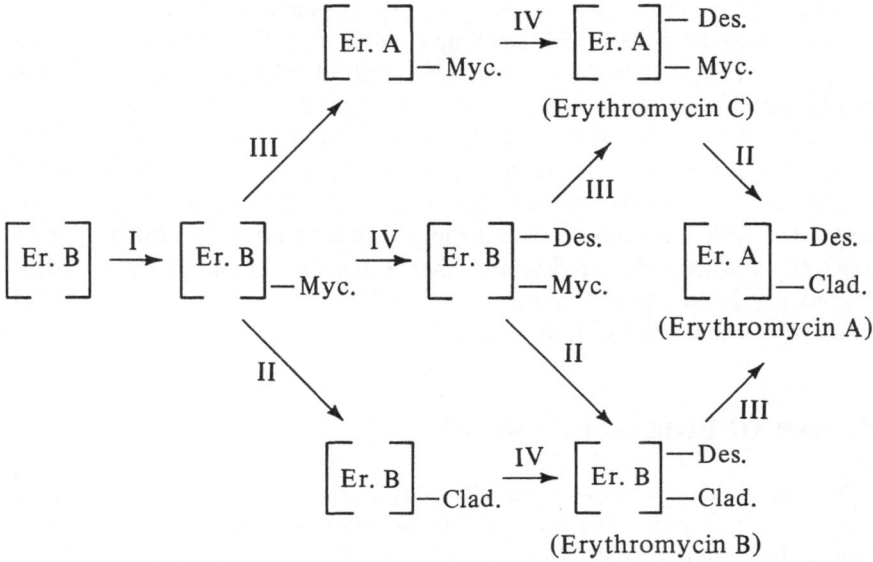

Figure 6.2. Example of metabolic net. Biosynthesis of erythromycins. (See text for explanation. Structures are shown in Fig. 6.22.)

the final molecule. The "starting point" for the study is thus the identification of those building blocks from which the molecule is made. For this initial aspect, tracer techniques are generally used. The possible precursors of the antibiotic, labeled with an isotopes such as ^{14}C, ^{3}H, or ^{13}C, are added to cultures of the producing organism, preferably at the time of maximum antibiotic production. At the end of the fermentation the antibiotic is extracted, purified, and the incorporation of the tracer isotope determined. When a radioactive label is used, the simple counting of radioactivity gives indications about the extent of incorporation of the precursor. However, the results must be interpreted very carefully because: (1) The precursor could fail to be taken up by the cells, generating a "false negative." (2) The precursor could be degraded to simpler molecules and the fragments incorporated through general metabolic pathways (false positive). It is therefore important to use precursors specifically labeled in one or more atoms and verify, through degradation of the final molecule, whether the radioactivity has been specifically incorporated and in which atoms.

The chemical degradation of fairly large molecules into defined fragments is often difficult and time consuming. Thus, alternative techniques to the radioactive labeling have been developed. The most important of these techniques involves the use of ^{13}C as the label. This carbon isotope has a natural abundance of about 1.3%. By nuclear magnetic resonance (NMR) it is possible to determine the ^{13}C component of each carbon atom in an organic molecule. The available instruments show it in the form of a resonance spectrum in which each carbon is represented by a peak. Precursors whose carbons are enriched up to 90% with ^{13}C can be synthesized (many are now commercially available) and added to the cultures of the producing strains. After completion of fermentation and isolation of the antibiotic, the incorporation of the label into the carbons of the final molecule is revealed as an increase in the height of the corresponding peaks in the magnetic resonance spectrum. The main difficulty that may be encountered with this method is that the precursor cannot be added in trace amount as with radioactivity, but that substrate quantities are needed. Moreover, the interpretation of a complex NMR spectrum requires special skill, experience, and substantial experimental work.

B. Use of Blocked Mutants

When the precursors have been identified it is generally possible to make a reasonable hypothesis regarding the biosynthetic pathway. It is important at this point to identify at least some of the intermediate in the sequence of reactions through which the final molecule is made.

This can be done by submitting the producing organism to mutagenic treatment and selecting the mutant strains that have lost the producing ability. When this is due to a single mutation that inactivates one of the enzymes involved in the biosynthetic pathway, the strain is called a blocked mutant. The product that is the substrate for the blocked enzymatic reaction cannot be further transformed and may accumulate in the medium. It can then be isolated and identified. To prove that the product is really an intermediate in the biosynthetic pathway, it is necessary to demonstrate the capacity of the original strain to transform it into the final product. This can be accomplished by radiotracer techniques or sometimes by determining analytically its conversion by suspensions of cells of the parent strain in the absence of nutrients.

By mutagenic treatment many nonproducing strains can easily be isolated. To facilitate the identification of those that accumulate biosynthetic intermediates in the fermentation broth, the cosynthesis method is used. This consists of growing the mutants two at a time in one flask. Strains that do not produce when grown singly but do produce when grown together are mutants blocked at two different points of the biosynthetic pathway; the inability of one to synthesize an intermediate is compensated by the ability of the other to accumulate it.

C. Enzyme Identification

The use of tracer techniques and identification of the intermediates generally are sufficient to determine the sequence of reactions through which the antibiotic is produced. This sequence, however, cannot be considered proven until it is shown that the microorganism possesses the enzymes able to catalyze the single reactions. This aspect is studied with the usual techniques of biochemistry. The enzymatic activity is sought in cell-free systems and when possible the enzyme is purified and its properties determined. In relation to the production of families of antibiotics, it is often of great interest to determine the substrate specificity of the enzyme. Often the presence or absence of the enzyme in the producing and nonproducing variants of the microorganism is determined to confirm its role. Obviously the sequence of studies described here is an ideal one. In practice, it can happen that the clue to a biosynthetic pattern is given by the identification of a peculiar enzymatic activity before any intermediate compound is isolated, or that intermediates are isolated very easily, as a consequence of mutagenic treatment made to improve production yield. However, all the complex biosynthetic pathways elucidated in the last years have been demonstrated by the combined use of the techniques summarized above.

III. Antibiotics Derived from a Single Primary Metabolite

Many antibiotics exist that have structures that are obviously derived from an amino acid or from an intermediate in amino acid biosynthesis.

Even though some of these products have some importance, the biosynthesis of only a very few of them has been studied; therefore, we cannot make any general statements. Generalizations are also precluded by the variety of structures and biosynthetic pathways for the amino acids. We will describe the biosynthesis of one example of this class, chloramphenicol, one of the few that has been studied sufficiently.

The nucleoside antibiotics are rather more homogeneous. They are structural analogs of the normal nucleotides that make up the nucleic acids. Their derivation from nucleotides has been demonstrated in several cases and we will illustrate some examples.

A. Chloramphenicol

Secondary metabolites with aromatic structure can be made by two biosynthetic pathways: cyclization of chains made from acetate units (or more exactly, from malonate), which is discussed later, or through the normal biosynthetic route for primary aromatic metabolites, sometimes called the shikimic acid pathway from one of the characteristic intermediates. Chloramphenicol is made through this last pathway. The first product in the biosynthetic line that has been determined is p-aminophenylalanine, which is not a normal product of primary metabolism. Experiments with labeled precursors have shown that it is derived by the same

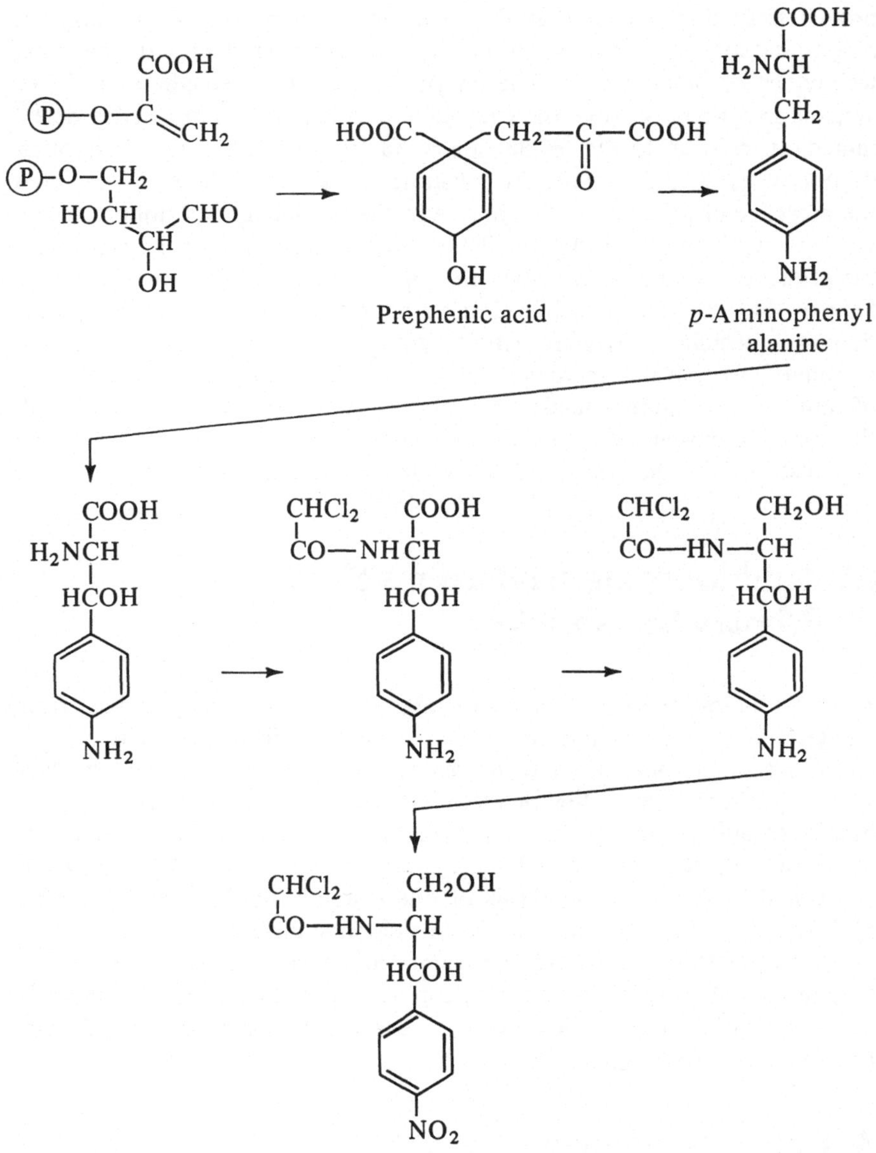

Chloramphenicol

Figure 6.3. Biosynthesis of chloramphenicol.

pathway as the aromatic amino acids. The last intermediate common to both pathways is probably prephenic acid. The succeeding intermediates in the synthesis have also been identified by experiments with labeled precursors and the overall biosynthetic scheme is shown in Fig. 6.3.

In this scheme, the interesting features are the formation of a primary alcohol group by reduction of a carboxyl and the origin of the nitro group, which confirms the notion that oxidized nitrogen functions (hydroxylamine, nitroso and nitro groups) are obtained by oxidation of amines.

B. Nucleoside Antibiotics

There are several antibiotics with structures analogous to those of natural nucleosides, but none are of much importance for antibacterial therapy. Their structures may differ from those of natural nucleosides by changes in the purine or pyrimidine base or in the sugar (ribose), and only very occasionally in both. Many of these antibiotics are synthesized from the normal nucleotide. One example is cordicepin or 3'-deoxyadenosine, derived from adenosine (or from adenylic acid) by the reaction shown in Fig. 6.4.

As shown in the figure, this is a simple reduction of a hydroxyl, and the skeleton of the molecule is not changed. Similar reactions occur in primary metabolism in the biosynthesis of 2'-deoxynucleotides.

More complex examples, with modification of the purine base, are the antibiotics tubercidin, sangivamycin, and toyocamycin, made from GTP through a series of reactions that have not yet been completely clarified, but that can be outlined as in Fig. 6.5.

The final products still contain the atoms of the starting material except for C-8 and nitrogen 9, which have been replaced by C-1 and -2 of a ribose molecule.

This type of reaction brings to mind the pathway in primary metabolism that leads to pteridine synthesis.

IV. Antibiotics Derived by Condensation of Several Metabolites

There are many antibiotics whose synthesis involves condensation of two or more molecules made in primary metabolism, which may have been more or less modified. Among those used in therapy are lincomycin and novobiocin, which are presented to illustrate this type of pathway.

1. Lincomycin

This molecule is obtained by condensation of a cyclic amino acid with a modified carbohydrate, but not much is known about the biosynthesis of the individual components. It has been shown that the methyl groups on the nitrogen, the sulfur, and the propyl chain are obtained from methionine.

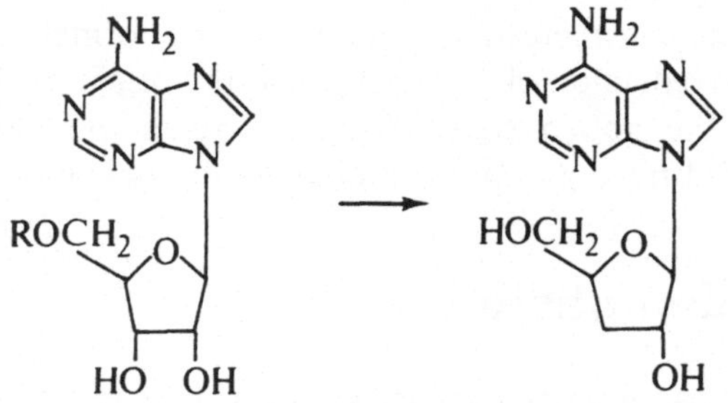

Figure 6.4. Biosynthesis of cordycepin.

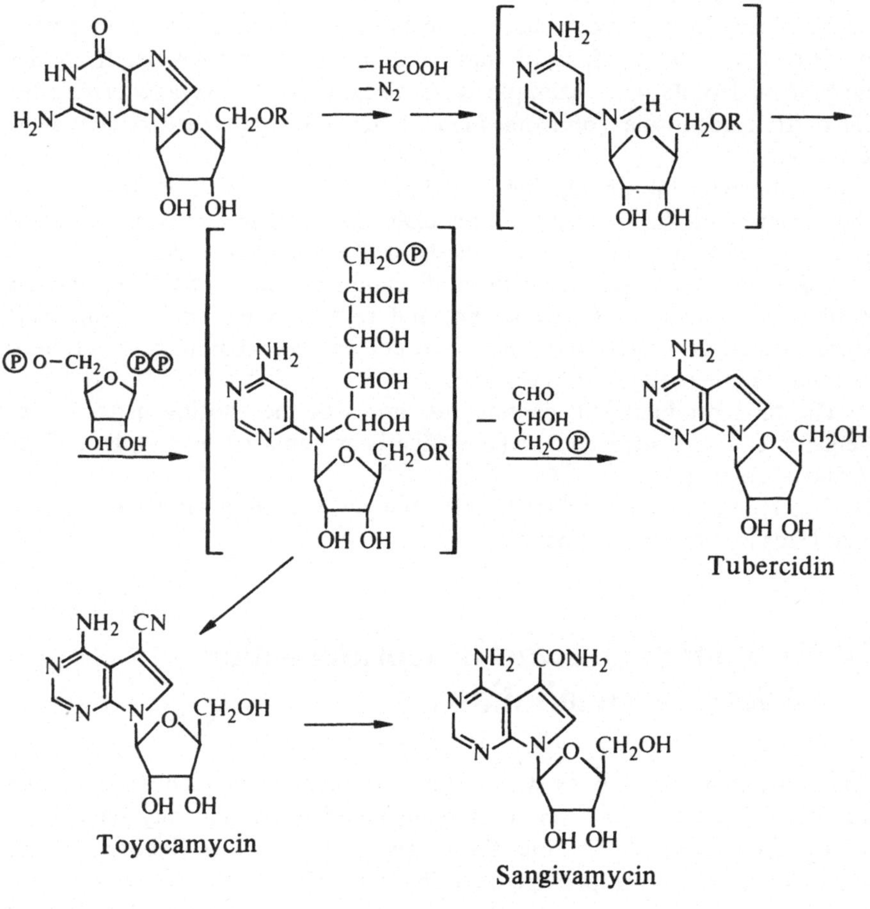

Figure 6.5. Biosynthesis of some nucleoside antibiotics.

The pyrrolidine ring is made from tyrosine by a degradative pathway that may parallel that by which melanin is made.

A partial and largely hypothetical scheme for the biosynthesis of lincomycin is given in Fig. 6.6.

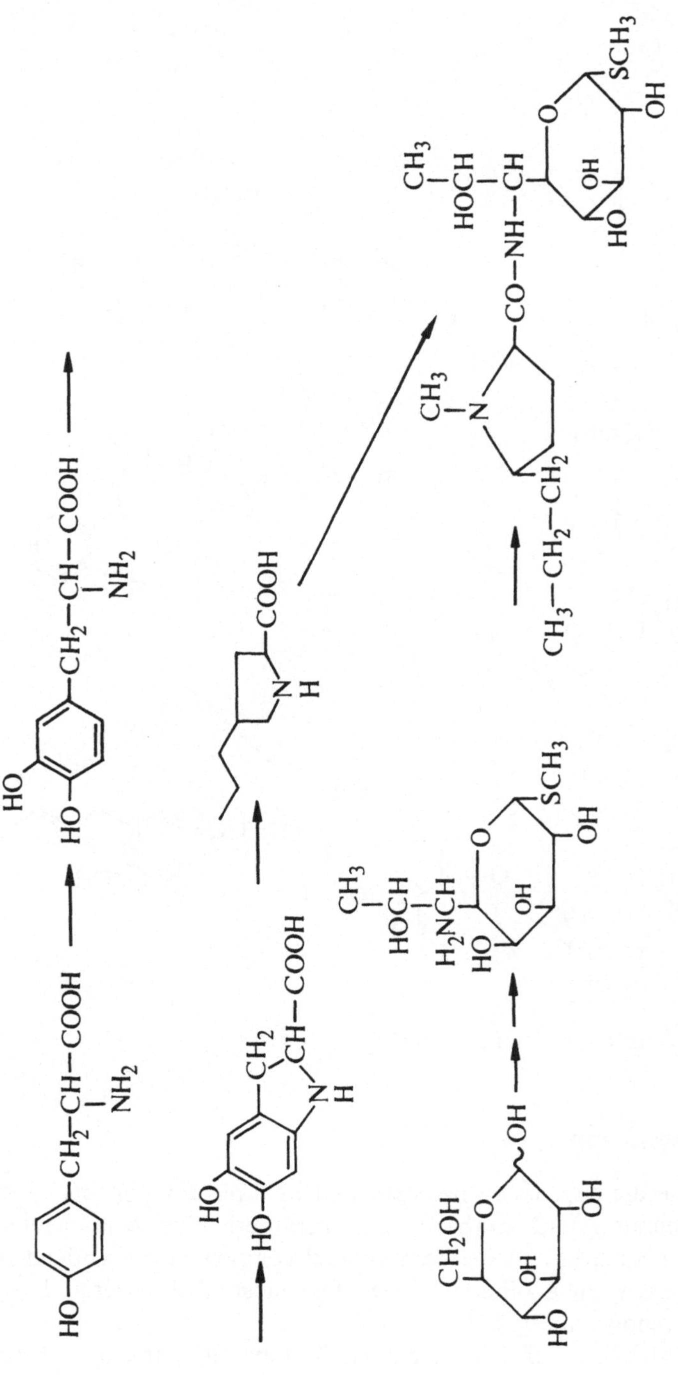

Figure 6.6. Hypothetical biosynthetic pathway for lincomycin.

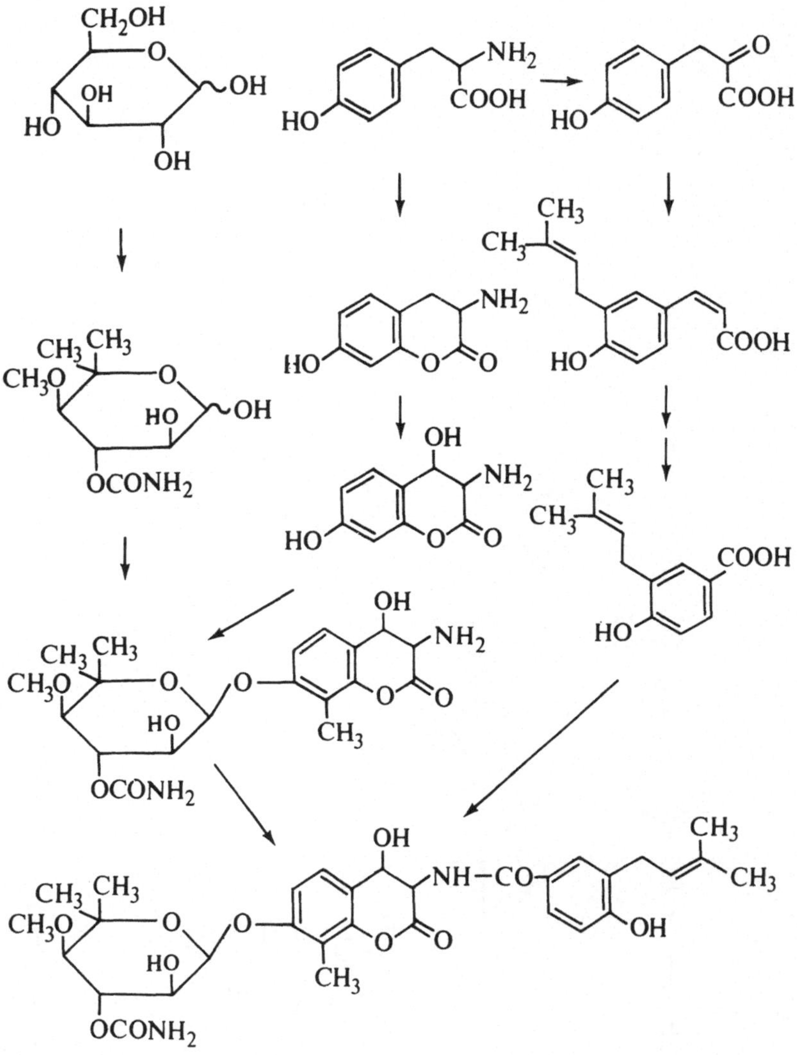

Figure 6.7. Biosynthesis of novobiocin.

2. Novobiocin

This molecule appears to be comprised of three components: a carbohydrate, coumarin, and another benzoic acid derivative. It is of interest that both the coumarin and the benzoic acid component are derived from the same primary metabolite, tyrosine. The most likely overall biosynthetic scheme is shown in Fig. 6.7.

The order in which these reactions is shown is partly hypothetical. For simplicity, we have shown the modifications of the sugar as taking place before condensation with the coumarin, but it is possible that the methylation and the carbamylation take place afterward.

We are also not certain at which stage the introduction of the aliphatic chain in position 3 of the benzene ring takes place. It is interesting that this chain is not derived from mevalonic acid, as one would suppose from its structure, and thus its biosynthesis is not isoprenoid.

V. Antibiotics Derived by Oligomer or Polymer Formation

Biosynthesis of antibiotics with a basal structure obtained by condensation of similar monomers can be divided into two phases: (1) the oligomerization process, which is similar for all the antibiotics of the same biosynthetic group and (2) terminal modifications characteristic of the individual antibiotics. This does not apply to the aminoglycoside antibiotics, in which the condensation of the individual units is usually the last phase in the biosynthesis.

A. Polypeptides and Depsipeptides

There are about 300 antibiotics that can be considered to originate from peptide condensation of amino acids.

They differ from normal proteins in the following characteristics:

1. Molecular weights of <3000 daltons;
2. the presence of uncommon amino acids, such as D-amino acids, N-methyl or hydroxyl amino acids, β-amino acids, etc.;
3. a tendency to cyclization and hypercyclization.

Some amino acids give rise to rings (for example, cysteine gives thiazoles) and frequently the molecule is a macrocycle.

Although it was disbelieved for a long time, the principal characteristic of the biogenesis of these products is that synthesis occurs through a specific enzyme complex. It was found that the normal protein synthesizing system of RNA-messenger-ribosome is not used. The first arguments in favor of this thesis were the presence in the peptide antibiotics of amino acids that are different from those coded for by the genetic code and the demonstration that their biosynthesis is not hindered by the presence of known inhibitors of protein synthesis. More recently, definitive proof was obtained when some of these antibiotics were synthesized in cell-free systems that contained neither ribosomes nor RNA.

These in vitro systems made it possible to identify the essential steps in the process of polymerization:

1. The amino acids are activated by reaction with ATP to give aminoacyl-adenylates.

2. The high-energy bond between the amino acid and the phosphate of the adenylic acid enables the amino acid to be transferred to thiol groups in a complex enzyme system, with the formation of amino-acylthio esters.

3. The enzyme system then catalyzes the formation of a peptide bond between the carboxyl group of the first amino acid and the amine of the second amino acid, using the energy obtained by breaking the thioester bond. The chain formed in this way is still esterified to a thiol group and can react with a third amino acid and so on to the end of the chain.

A more detailed description is given later for synthesis of gramicidin S.

It is important to note that this synthetic mechanism resembles that for the polymerization of malonate units in the synthesis of fatty acids. The principal difference consists of the fact that in the polypeptide synthesis system the sequence of amino acids is regulated, whereas in the synthesis of fatty acids all the monomers are identical and this type of regulation is not required.

This synthetic mechanism explains why (1) the peptide antibiotics are never longer than 15 or 20 amino acids, since it is unlikely that an enzyme system complex could "recognize" more than that, and (2) the peptide antibiotics are often produced in "families," the components of which differ in one or more amino acids. The system for recognition of the various amino acids can, in fact, be ambiguous and different amino acids can compete for the same position.

Very often, after or during the formation of the chain, further reactions occur. These can include cyclization of the molecule or of part of it and modification of individual amino acids. For instance, D-amino acids are often found in polypeptides usually made from L-amino acids and the inversion has been shown to occur after the polymerization. One example of a profound modification of a tripeptide chain is in the biosynthesis of penicillins and cephalosporins, which are described later.

It is probable that similar mechanisms of biosynthesis operate in the biogenesis of peptolides and depsipeptides, antibiotics that differ from polypeptides in having some ester bonds instead of amide bonds.

1. Biosynthesis of Gramicidin S

Gramicidin S is a decapeptide composed of two identical pentapeptides cyclized head-to-tail (Fig. 6.8).

Biosynthesis of this antibiotic is carried out by two soluble enzymes, one with a molecular weight of 100,000 (the light enzyme, or L) and one with a molecular weight of 280,000 (the heavy enzyme, or H). Extracts of the producing strain, *Bacillus brevis,* that contain these enzymes

Leu — — — — — → Phe

↑   ↓

Orn   Pro

↑   ↓

Val   Val

↑   ↓

Pro   Orn

↑   ↓

Phe ← — — — — — Leu

Figure 6.8. Structure of gramicidin S. The *solid arrows* indicate —CO—NH— bonds. The *dashed arrows* indicate bonds forming the cycle of two subunits. Phe, phenylalanine; Pro, proline; Val, valine; Orn, ornithine; Leu, leucine.

can synthesize the antibiotic from the amino acids plus ATP and magnesium ions. The light enzyme catalyzes the activation of phenylalanine and its conversion from levorotatory to dextrorotatory; the heavy enzyme catalyzes the activation of the other four amino acids. Activation takes place by the reactions that are outlined in Fig. 6.9.

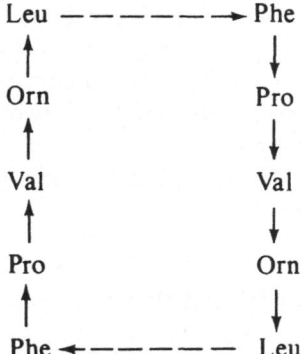

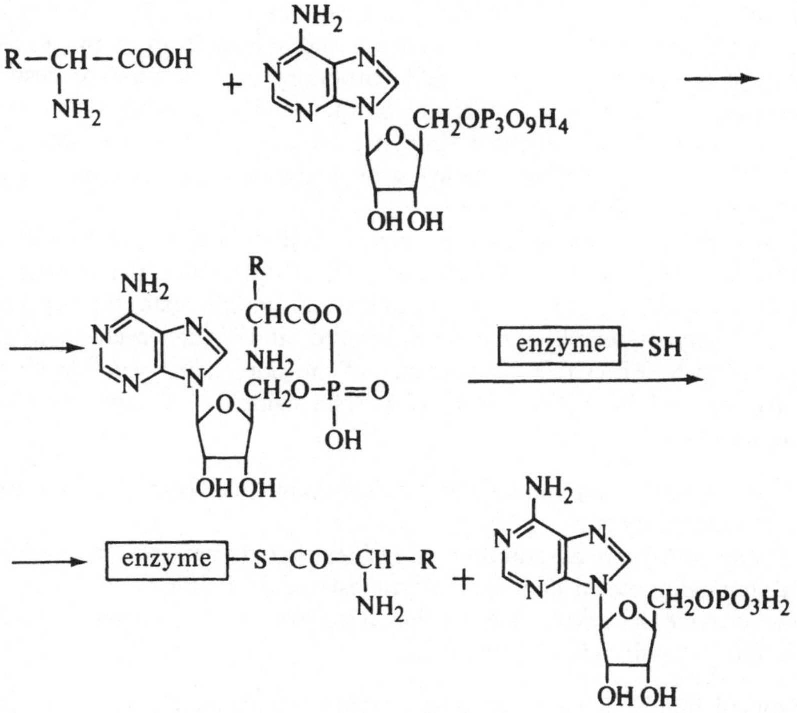

Figure 6.9. Activation of amino acids during biosynthesis of gramicidin S.

After activation, the five amino acids are bound as thioesters to the enzymes. Polymerization takes place only when the two enzymes that carry the activated amino acids are joined, and occurs as shown in Fig. 6.10.

Once the pentapeptide has been completed, two molecules react to form a cyclic compound closed by two additional leucine-phenylalanine peptide bonds, with cleavage of the last thioester bonds and liberation of the enzyme.

2. Penicillins and Cephalosporins

The biosynthesis of these antibiotics has been studied mostly in *Penicillium chrysogenum* and *Cephalosporium acremonium*. *P. chrysogenum* produces different penicillins, depending on the fermentation conditions. Some which are important from the biosynthetic point of view are shown in Fig. 6.11.

C. acremonium produces penicillin N and cephalosporin C, shown in Fig. 6.12, and a third antibiotic with a steroid structure.

The biosynthesis of the side chain of penicillin G from phenylacetic acid has been shown by (1) addition of phenylacetic acid to culture of *P. notatum,* which causes an increase in penicillin G production, and (2) addition of radioactively labeled phenylacetic acid, which leads to specific labeling of the side chain of the penicillin.

Similar experiments with labeled precursors have shown that the nucleus of the penicillins is derived from cysteine and valine. Both of these molecules are incorporated intact. The levorotatory configuration of cysteine is unchanged. There is an inversion of the configuration of valine from levo- to dextrorotatory at the asymmetric center but it has been shown by labeling separately the two methyl groups that the configuration of the third carbon atom is not changed.

However, it is unlikely that the microorganism directly synthesizes the penicillin nucleus (6-aminopenicillanic acid, 6-APA) and then condenses it with phenylacetic acid. Rather, an obligatory intermediate, the tripeptide α-aminoadipyl-cysteinyl-valine, is synthesized, and is then cyclized to give isopenicillin N. From this, by exchange of the side chain, penicillin G can be derived and by hydrolysis 6-APA. The evidence in support of this hypothesis is:

1. The tripeptide α-aminoadipyl-cysteinyl-valine has been isolated from fermentation broths.
2. Conditions, such as addition of lysine, that depress the synthesis of α-aminoadipic acid depress the synthesis of penicillin G.
3. An enzyme, an acyltransferase, has been isolated from *Penicillium* that is able to catalyze in vitro the reactions:

Isopenicillin N + phenylacetyl coenzyme A → penicillin G + α-aminoadipic acid + coenzyme A

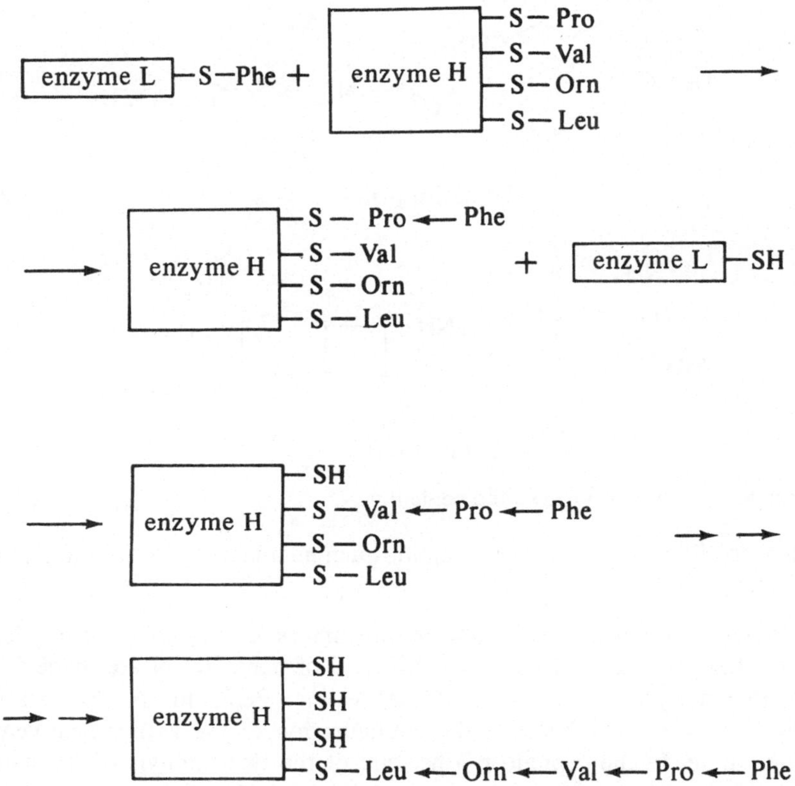

Figure 6.10. Biosynthesis of gramacidin S.

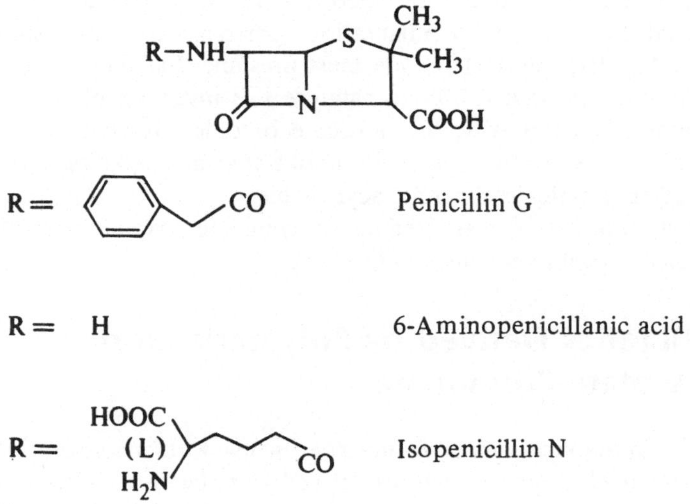

Figure 6.11. Penicillin and its derivatives.

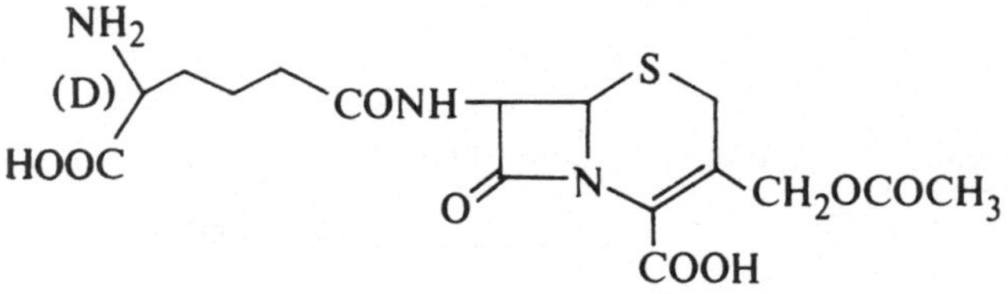

Cephalosporin C

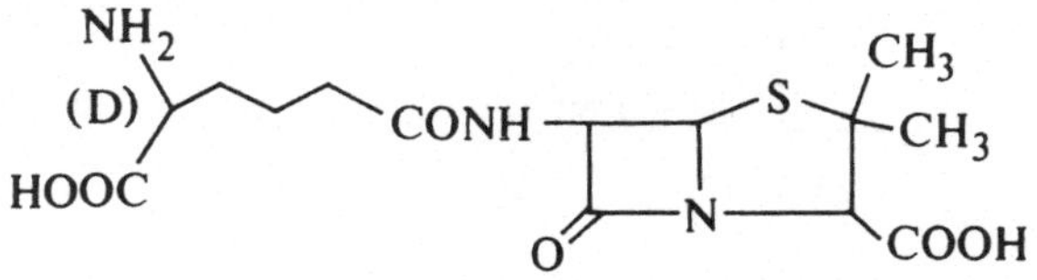

Penicillin N

Figure 6.12. Cephalosporin C and penicillin N.

Isopenicillin N + H$_2$O → 6-amino-penicillanic acid + α-aminoadipic acid.

Similar experiments with labeled precursors in *Cephalosporium* have shown that the nuclei of cephalosporin and of penicillin N are made from levorotatory cysteine and levorotatory valine. Penicillin N, like all the penicillins, shows inversion of the configuration of the asymmetric center of the valine. In the cephalosporins, one of the two methyls of the valine is incorporated into the dehydrothiazine ring; the other is first hydroxylated and then acetylated. The derivation of cephalosporin C and penicillin N from the tripeptide α-aminoadipyl-cysteinyl-valine is fairly obvious from their structure. Furthermore, this tripeptide has been isolated from *Cephalosporium*. However, it should be noted that in the tripeptide the α-aminoadipic acid has the levo configuration, whereas in cephalosporin and penicillin N it has the dextro. One must presume that there is an enzyme in *Cephalosporium* that is able to catalyze this inversion of configuration. Once modified in this way, the molecule becomes resistant to acyltransferases, and therefore the side chain, unlike that in *Penicillium,* cannot be exchanged enzymatically for other acyl chains.

Current information was used in showing the overall biosynthesis of penicillins and cephalosporins in Fig. 6.13.

B. Antibiotics Derived by Polymerization of Acetate-Propionate

Numerous antibiotics appear from experiments with incorporation of labeled acetic acid to have structures derived from head-to-tail condensation of acetate units, i.e., with bond formation between the carbon of the car-

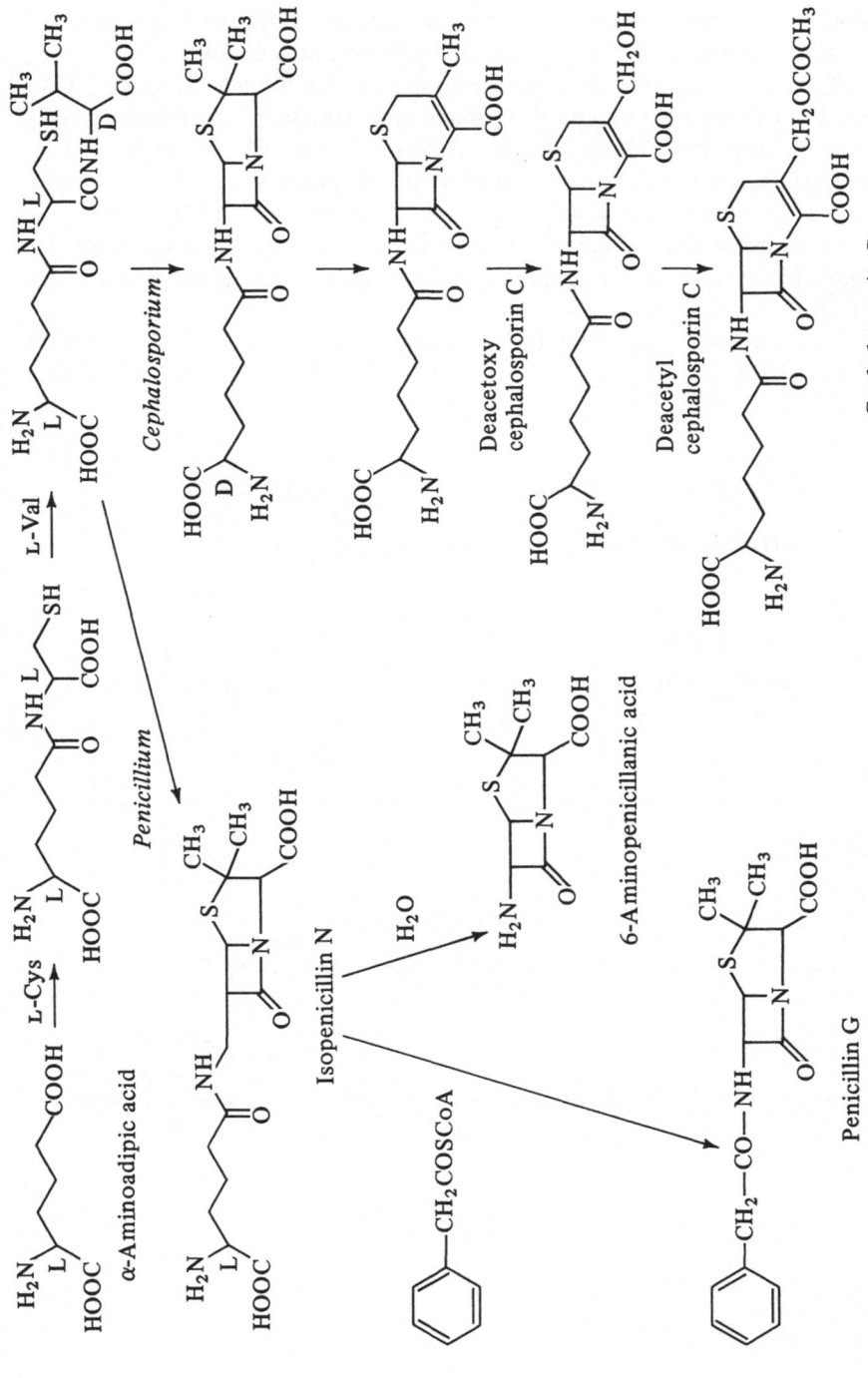

Figure 6.13. Biosynthesis of penicillin and cephalosporin.

boxyl and that of the methyl carbon of the next unit. Sometimes the structural unit appears to have been propionate instead of acetate, with the bond in this case formed between the carbon of the carboxyl and the methylene of the next unit, giving a methyl-substituted chain.

Quite different structures can be found in this biogenetic group. They are the isolated or condensed aromatic ring structures, quinones, macrolides, ansamycins. These marked differences are the consequences of relatively modest variations in the biosynthetic processes, which are fundamentally similar in many ways to those giving rise to the fatty acids.

To examine the polymerization process by which the fatty acids are made, let us take as an example the biosynthesis of palmitic acid (Fig. 6.14).

It can be seen that acetic acid is directly involved only in the initiation of the chain, whereas malonic acid is the basal unit for the elongation

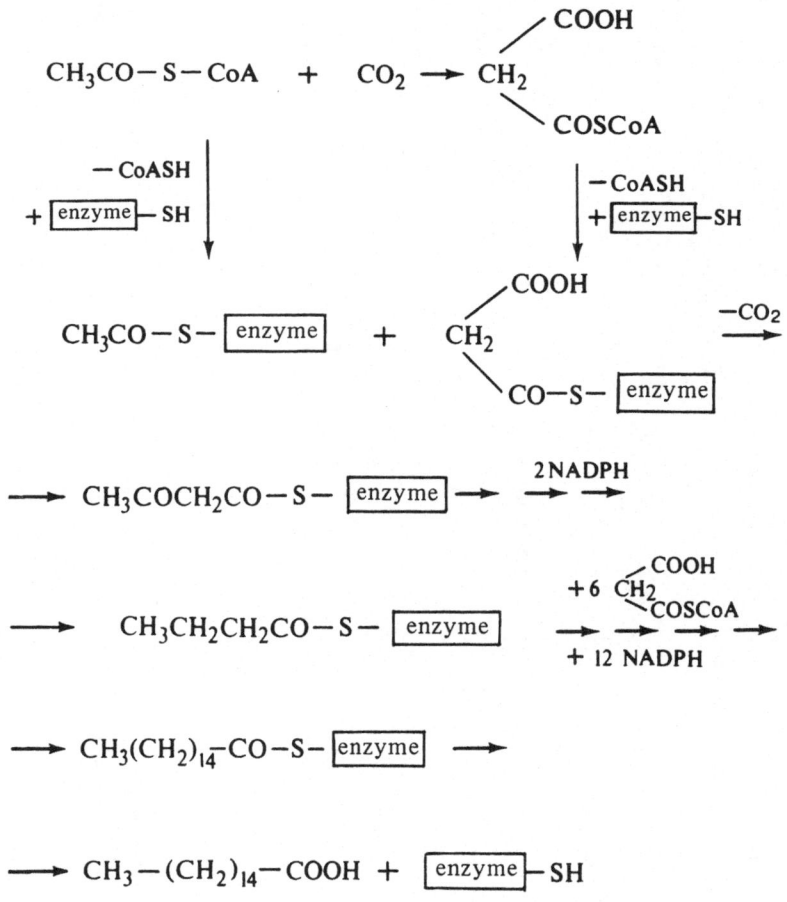

Figure 6.14. Biosynthesis of palmitic acid.

process. Malonic acid is, however, derived from acetic acid by carboxylation and for this reason, experiments with incorporation of labeled precursors indicate the chain is made entirely from acetate units.

For chain elongation to occur, both the acetic acid and the malonic acid must first be activated. Activation occurs through the formation of a thioester bond with coenzyme A (CoASH). Acetate and malonate are transferred through this coenzyme thioester bond to the enzyme system that carries out the polymerization, the reduction of the carbonyls to methylenes [by reduction to the hydroxy, dehydration, and reduction of the double bond (not shown in the outline)]. In eukaryotic cells these reactions are carried out by a multienzyme complex of very large molecular weight ($>2,000,000$), whereas in bacteria there are separate enzymes for the various reactions.

The principal difference between the system that synthesizes fatty acids and that giving rise to the secondary metabolites lies in the reduction of the carbonyl groups, which occurs in fatty acid synthesis immediately after condensation of a unit and before proceeding to the next condensation. In the biosynthesis of the secondary metabolites, this reduction either does not occur, giving rise, as we will see, to aromatic structures, or takes place only partially, to the hydroxyl or double bond stage, giving macrolide structures.

Another important difference is the ability of the systems that synthesize some antibiotics to use methylmalonate (which may be obtained from carboxylation of propionate) for the elongation, with formation of methylated chains. In some cases ethylmalonate can also be used to elongate the chain, resulting in formation of ethyl-substituted chains. Just as incorporation of propionate indicates that the chain was synthesized from methylmalonate units, the incorporation of butyrate indicates that the intermediate was ethylmalonate.

The molecule that initiates the chain in fatty acid synthesis is usually activated acetic acid, but in fatty acid synthesis by bacteria there can be other initiators, such as propionate and isovaleric acid.

In the biosynthesis of the antibiotics, in addition to acetic and propionic acids, we can have as initiator malonamide and other more complex molecules, as we shall see later. An outline of the formation of the chain that gives rise to antibiotics derived from acetate and propionate is given in Fig. 6.15.

For the link between the primary and secondary metabolism of the cell, remember that acetylcoenzyme A is a normal product of glucose metabolism, malonylcoenzyme A is also a normal metabolic product synthesized by carboxylation of acetylcoenzyme A, and methylmalonylcoenzyme A can be derived analogously by carboxylation of propionylcoenzyme A, but is obtained more frequently through other pathways, such as, e.g., isomerization of succinylcoenzyme A.

The structure of the antibiotic is basically determined by the length of the chain and by the degree of reduction of the chain. Chains formed

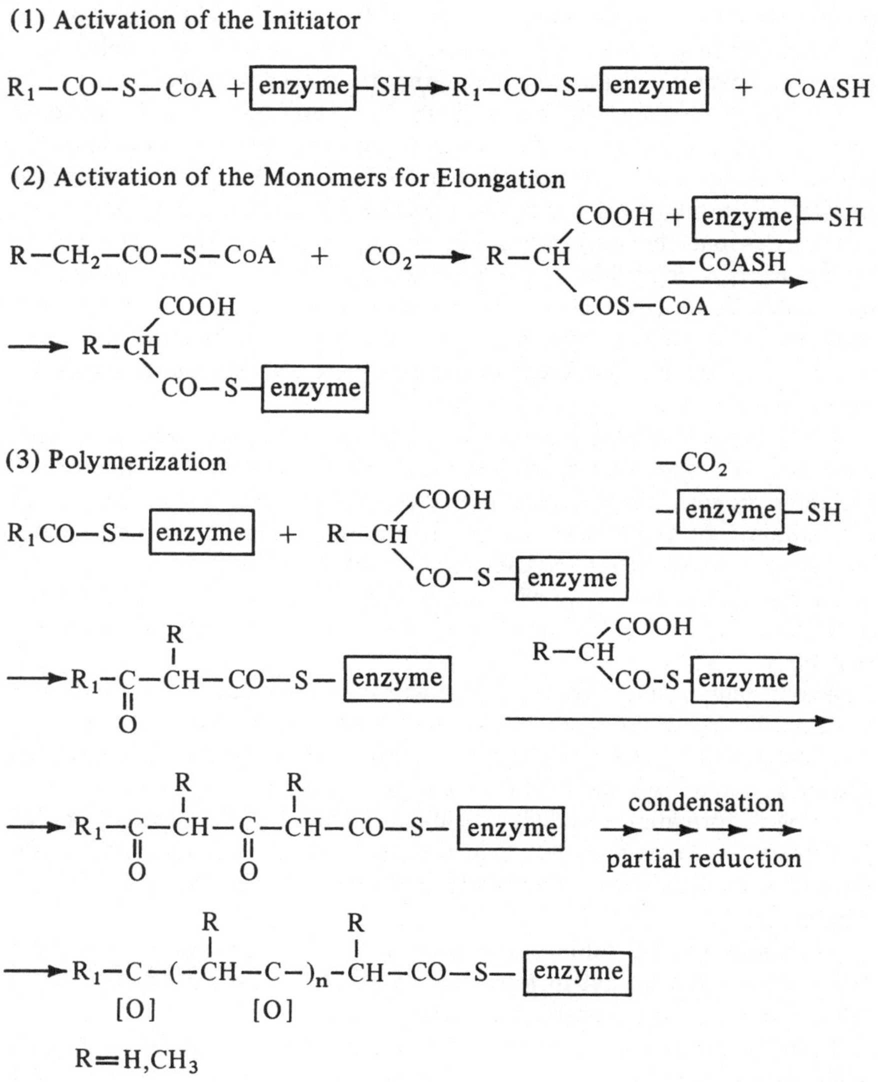

(1) Activation of the Initiator

$$R_1-CO-S-CoA + \boxed{enzyme}-SH \longrightarrow R_1-CO-S-\boxed{enzyme} + CoASH$$

(2) Activation of the Monomers for Elongation

$$R-CH_2-CO-S-CoA + CO_2 \longrightarrow R-CH \begin{smallmatrix} COOH \\ \\ COS-CoA \end{smallmatrix} \quad \begin{smallmatrix} +\boxed{enzyme}-SH \\ -CoASH \end{smallmatrix} \longrightarrow$$

$$\longrightarrow R-CH \begin{smallmatrix} COOH \\ \\ CO-S-\boxed{enzyme} \end{smallmatrix}$$

(3) Polymerization

$$R_1CO-S-\boxed{enzyme} + R-CH \begin{smallmatrix} COOH \\ \\ CO-S-\boxed{enzyme} \end{smallmatrix} \qquad \begin{smallmatrix} -CO_2 \\ -\boxed{enzyme}-SH \end{smallmatrix} \longrightarrow$$

$$\longrightarrow R_1-\overset{R}{\underset{O}{\overset{|}{C}}}-CH-CO-S-\boxed{enzyme} \qquad R-CH \begin{smallmatrix} COOH \\ \\ CO-S-\boxed{enzyme} \end{smallmatrix} \longrightarrow$$

$$\longrightarrow R_1-\overset{R}{\underset{O}{\overset{|}{C}}}-CH-\overset{R}{\underset{O}{\overset{|}{C}}}-CH-CO-S-\boxed{enzyme} \qquad \begin{smallmatrix} condensation \\ \rightarrow \rightarrow \rightarrow \rightarrow \\ partial\ reduction \end{smallmatrix}$$

$$\longrightarrow R_1-\overset{R}{\underset{[O]}{\overset{|}{C}}}-(-\overset{R}{\underset{[O]}{\overset{|}{CH}}}-C-)_n-CH-CO-S-\boxed{enzyme}$$

$$R=H, CH_3$$

Figure 6.15. Biogenesis of the chain that is the origin of antibiotics derived from acetate-propionate.

entirely from acetate, if the carbonyls are not reduced, consist of alternate methylene and carbonyl groups (polyketomethylene or polyketide chains). The methylenes in these structures are highly active and tend to condense with carbonyl groups to form rings, which for steric reasons tend to contain six atoms. Enolization of the carbonyls that have not reacted leads to aromatization of the structure. Some examples are shown in Fig. 6.16.

Whenever, because of partial reduction during the polymerization proc-

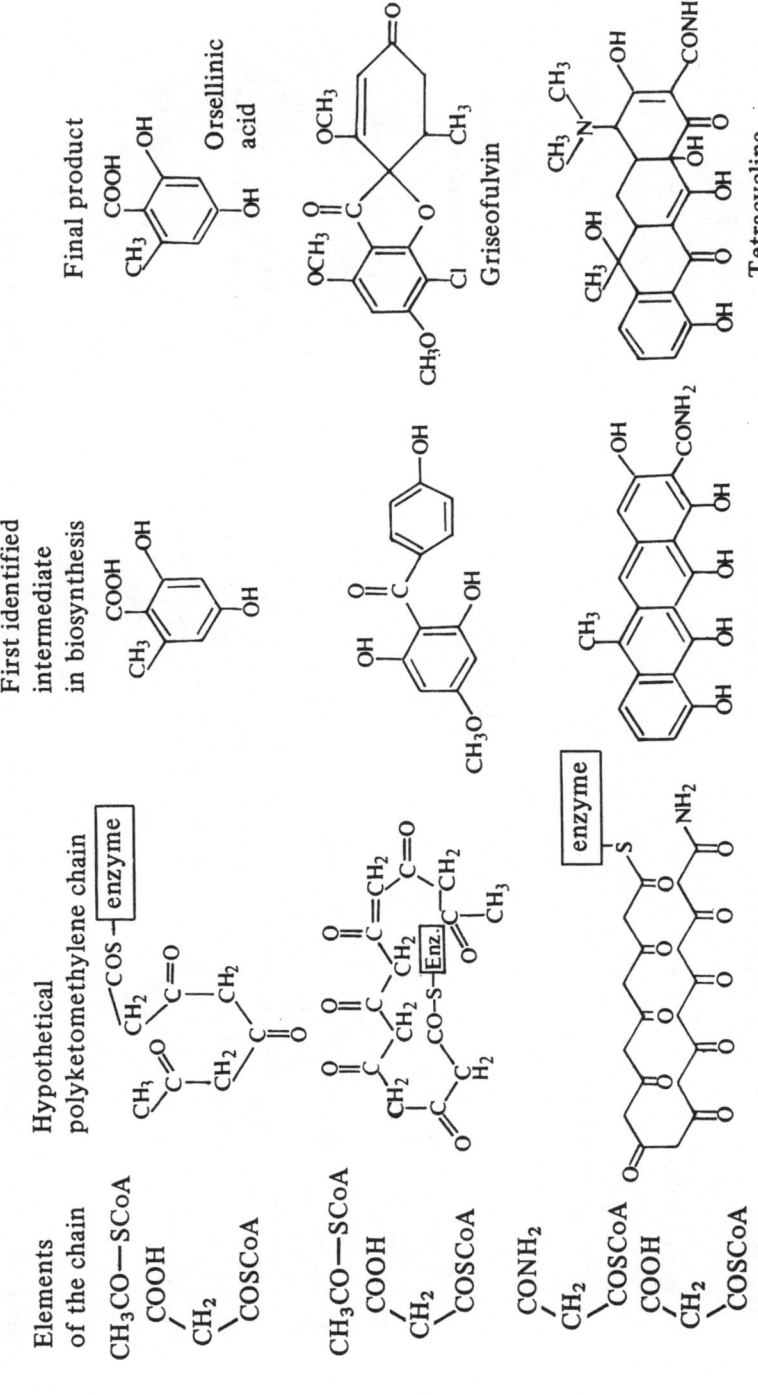

Figure 6.16. Formation of aromatic rings from polyketomethylene chains.

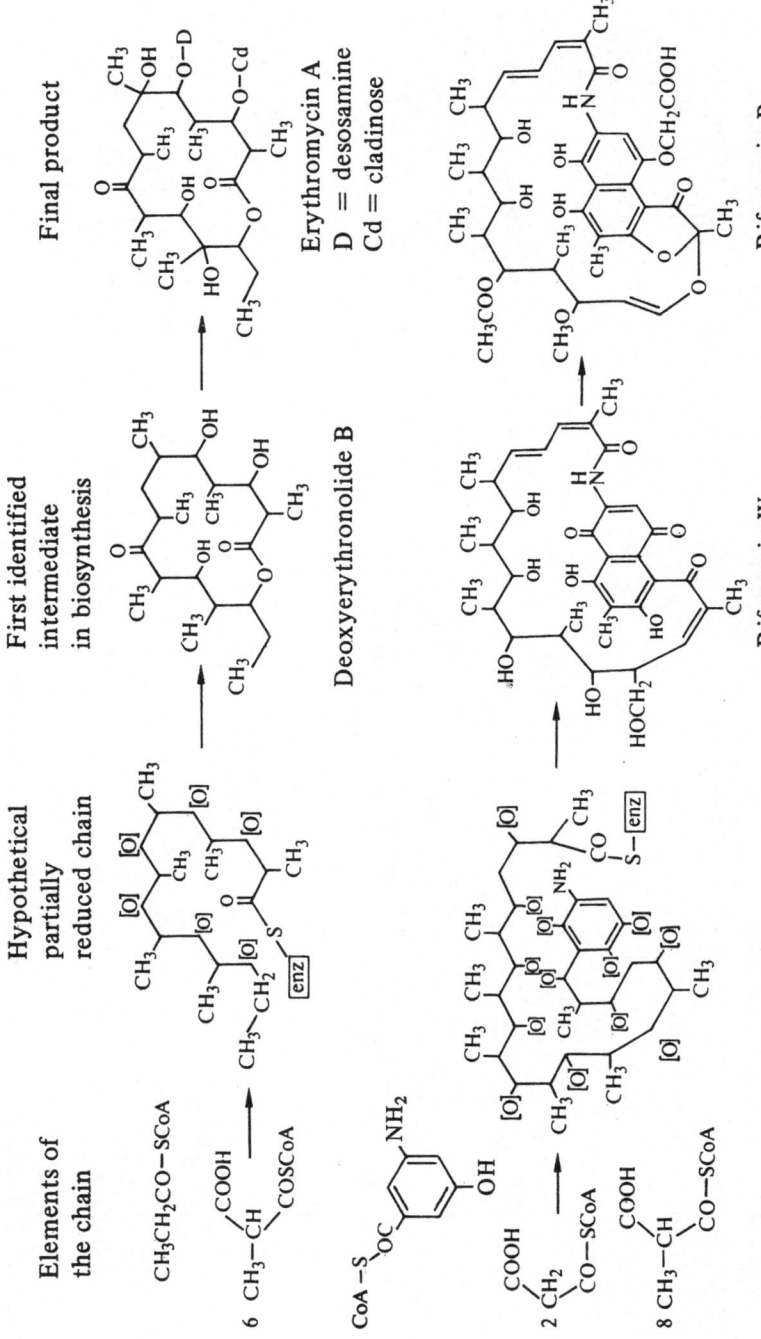

Figure 6.17. Formation of macrolides and ansamycins from partially reduced polyketomethylene chains.

ess or because methyl substituents are present, the chain is not truly polyketomethylene, cyclization and formation of aromatic rings is more difficult and one obtains linear structures or, more commonly, macrocycles. Classic examples of this are the macrolides, both the antibacterial types with medium-sized rings and the antifungal types with large rings (Fig. 6.17). The ansamycins may be considered to be in a way intermediate, with formation of both an aromatic ring and a macrocycle (Fig. 6.17).

We emphasized earlier that after formation of the basal structure usually other enzymatic reactions complete the synthesis of the antibiotic or the family of antibiotics. We shall discuss some of these transformations for some of the antibiotics that are most important or some of the biosyntheses that are most interesting.

1. Griseofulvin

The biosynthesis of this antibiotic in *Penicillium griseofulvum* and *Penicillium patulum* was studied both with addition of labeled precursors and of isolated metabolic products presumed to be intermediates. We stated previously that from one acetate and six malonate units a substituted benzophenone is formed. An outline of the probable successive reactions is indicated in Fig. 6.18.

It can be seen that one goes from a completely aromatic structure to a partially saturated one. This type of pathway is fairly common in biosyntheses from acetate.

2. Tetracyclines

The common precursor of tetracycline, chlortetracycline, and oxytetracycline is methylpretetramide (a), which is obtained by condensation of one malonamide unit with eight malonate units. The methyl comes from

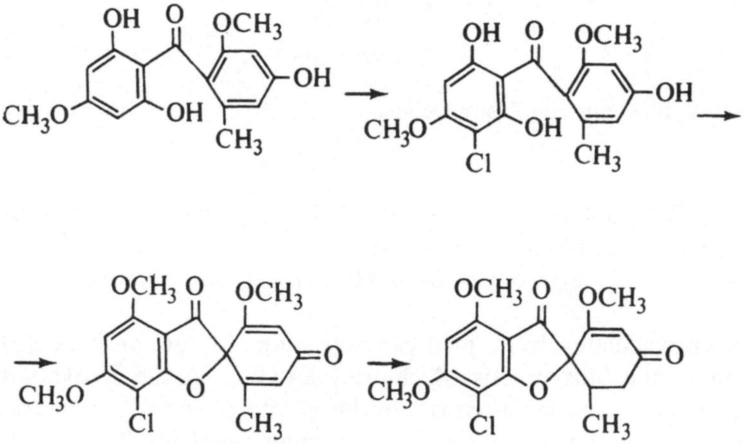

Figure 6.18. Final steps in biosynthesis of griseofulvin.

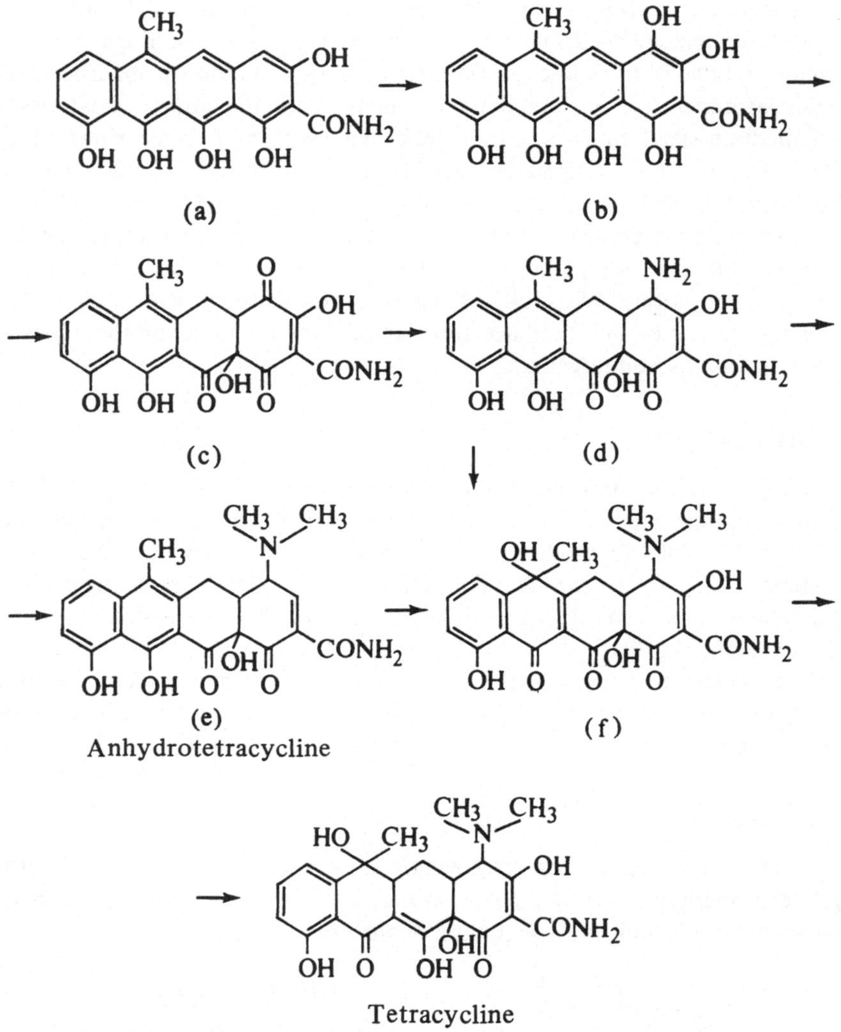

Figure 6.19. Biosynthesis of tetracycline.

the usual C-1 pool (that is, the methyl donor pool). The reactions that convert methylpretetramide into tetracycline, as were demonstrated by isolation of the intermediates from blocked mutants, are shown in Fig. 6.19.

The intermediates from methylpretetramide to the product (c) are common to the biosynthesis of chlortetracycline. In the biosynthesis of that antibiotic, a chlorine atom is introduced into position 7 of product (c) and the later reactions proceed with the chlorinated product as with the nonchlorinated one (Fig. 6.20).

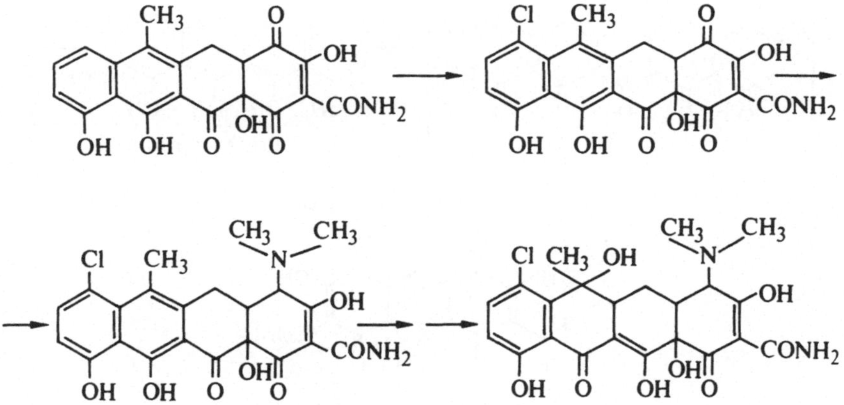

Figure 6.20. Biosynthesis of chlortetracycline.

The hydroxyl group in position 5 of oxytetracycline is introduced at a later stage, when anhydrotetracycline (d), is converted to anhydrooxytetracycline (Fig. 6.21). This molecule also undergoes the same subsequent reactions as in tetracycline biosynthesis.

If the fermentation is carried out in the presence of inhibitors of methylation, for example, sulfonamides, or with a mutant strain that cannot methylate, pretetramide is obtained instead of methylpretetramide (a) (Fig. 6.22). The biosynthesis proceeds in the same way as described

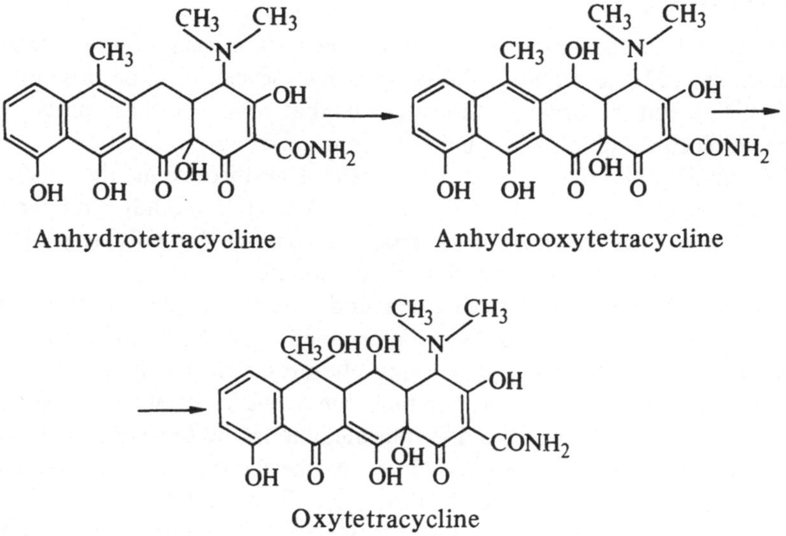

Anhydrotetracycline   Anhydrooxytetracycline

Oxytetracycline

Figure 6.21. Biosynthesis of oxytetracycline.

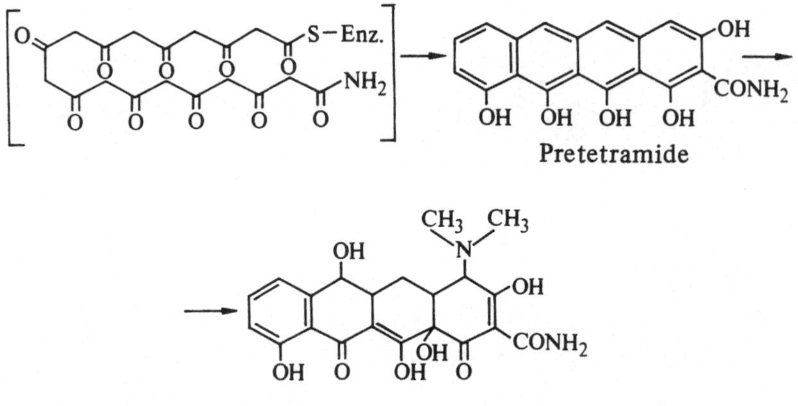

Figure 6.22. Biosynthesis of 6-demethyltetracycline.

before to give demethyltetracycline (or demethylchlortetracycline). It should be noted that methylation of the nitrogen takes place normally, indicating that this occurs through a mechanism that differs from that for methylation of the carbon.

The effects of primary metabolism on biosynthesis of secondary metabolites become apparent for the tetracyclines when the culture conditions or addition of substances favor the metabolism of glucose through the pentose pathway. This leads to an increase in production of the antibiotic, which is produced less when glucose is metabolized by glycolysis.

3. Erythromycin

There are three aspects to consider in the biosynthesis of the macrolide antibiotics: (1) the origin of the lactone macrocycle, (2) the origin of the sugars, (3) and the order of the reactions that condense these parts to the whole molecule and introduce modifications.

We shall summarize what is known about erythromycin, the macrolide that has been studied most in this regard. We have already said that the basic structure of the lactone macrocycle is made by polymerization of one unit of propionate with six of methylmalonate.

In the different erythromycins one finds the sugars desosamine, mycarose, and cladinose. It has been determined that the basic structure of these arises from glucose, without any alterations in the order of the carbon atoms. A possible outline for their biosynthesis is given in Fig. 6.23.

Deoxyerythronolide B, the first intermediate in the biosynthesis that has been isolated to date, is converted by hydroxylation to erythronolide B, which can be converted into erythromycin B by condensation with the desosamine and cladinose, or to erythromycin A by hydroxylation in posi-

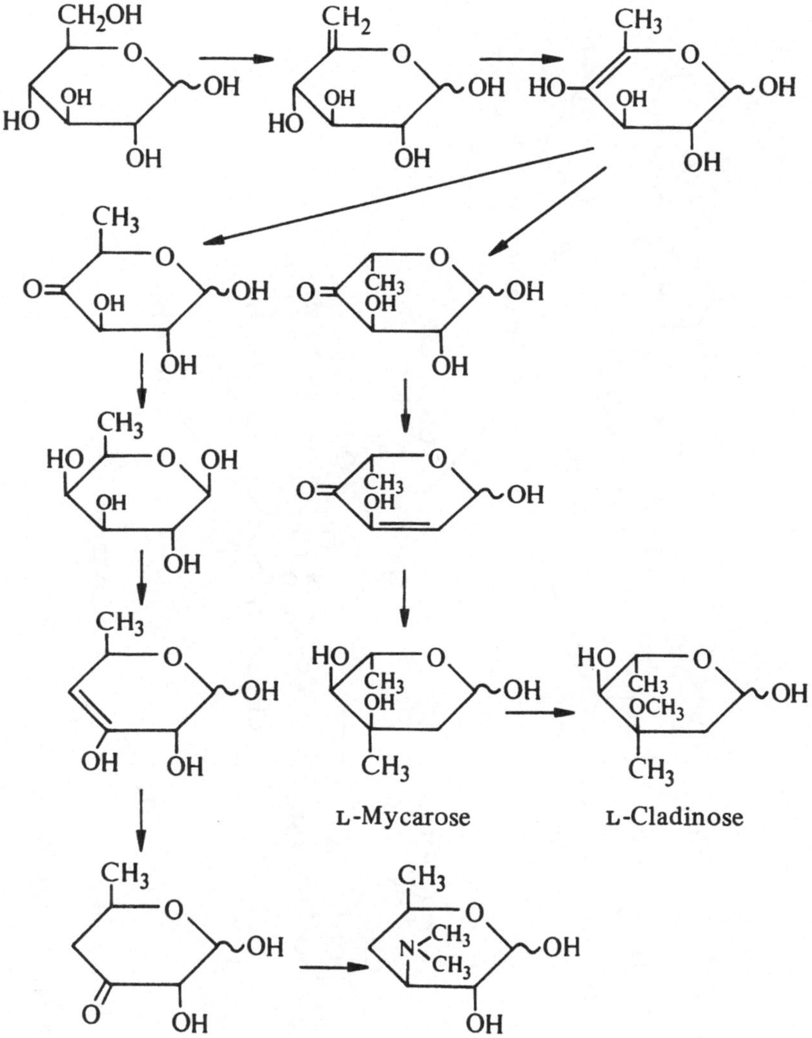

Figure 6.23. Biosynthesis of the sugars of erythromycin.

tion 12 and condensation with desosamine and cladinose, or into erythromycin C by hydroxylation in position 12 followed by condensation with
desosamine and mycarose.

There is probably no strict order for these reactions to take place, and
there is the chance of interconversions of some of the intermediate metabolites, so that the more complex product, erythromycin A, could be made
by more than one biosynthetic pathway, as indicated in Fig. 6.24.

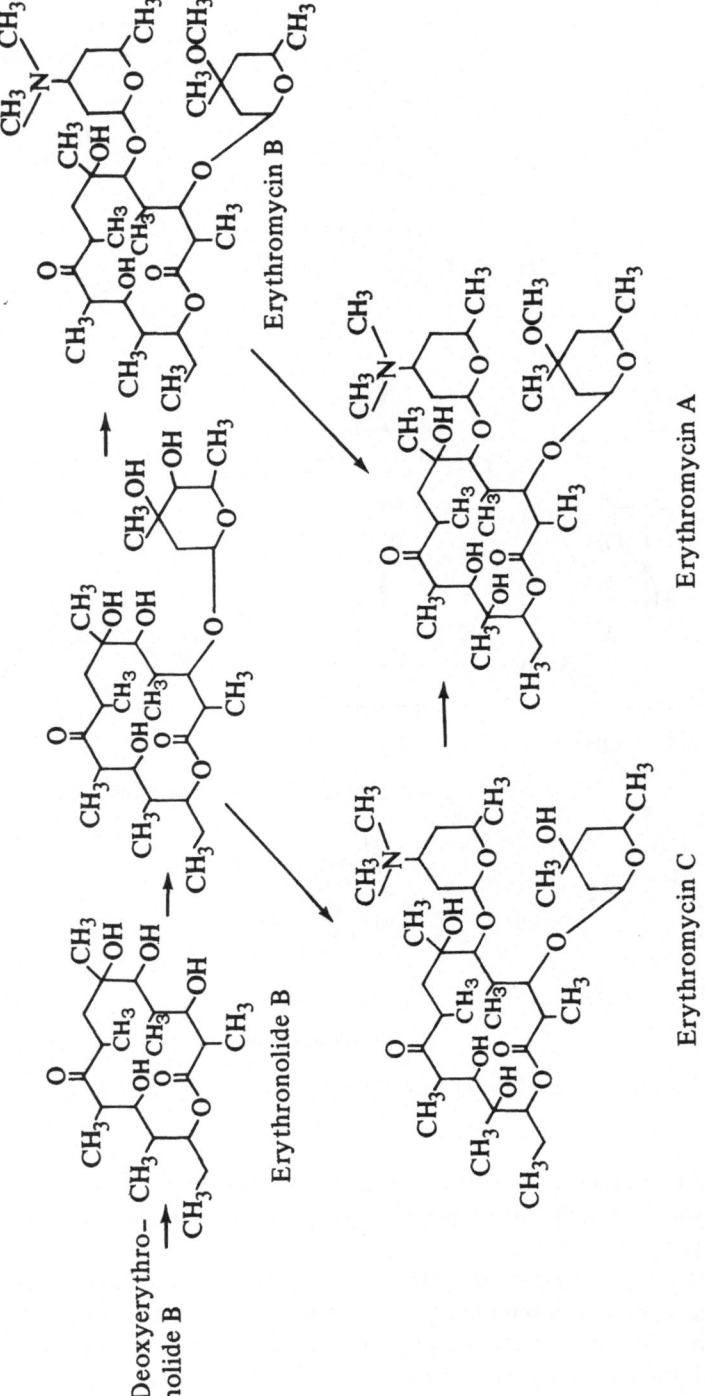

Figure 6.24. Biosynthetic relationships of the erythromycins.

4. Rifamycins

The basic structure of the molecule of the naphthalene ansamycins (rifamycin, streptovaricin, tolipomycin, halomycin) is derived from a chain formed by condensation of two malonate units and eight methylmalonate units.

Characteristic of the ansamycins is the group that initiates the chain, a compound recently shown to be 3-amino-5-hydroxybenzoic acid. This compound is made as a variant in the biosynthetic pathway for aromatic amino acids, as shown by experiments with incorporation of [^{13}C]glucose. The methylmalonate units may be made from propionic acid if this is added to the culture, but usually appear to be made by isomerization of succinic acid.

The characteristic aspects of the biosynthesis of rifamycins and other ansamycins are the partial cyclization of the chain, with formation of a second aromatic ring condensed with that of the initiator, and closure of the macrocycle by an amide linkage.

The first intermediate in the biosynthesis of rifamycins, protorifamycin and rifamycin W, have the structure described above. By later reactions, among them breaking the aliphatic chain and insertion of an ethereal oxygen, rifamycin W is converted into the final products as shown in Fig. 6.25. The chief intermediate in these transformations is rifamycin S.

Rifamycin B is partly converted into a derivative with an oxidized chain, rifamycin Y, which is microbiologically inactive. Depending on the fermentation conditions or on the strain used, rifamycin S can give rise not only to rifamycin B but also to other products, such as rifamycin O and rifamycin G, whose partial structures are shown in Fig. 6.26, and other products with unknown structures that are called collectively the rifamycin complex. Other ansamycins such as tolipomycin and halomycin are thought to be biosynthesized from rifamycin W, either through rifamycin S or through a similar intermediate.

C. Terpenoid Antibiotics

Among the secondary metabolites of fungi, there are many products with terpenoid structures made by condensation of isoprene units (or, to be more precise, isopentenylpyrophosphate units), often followed by cyclization. The isopentenylpyrophosphate units are synthesized from acetate through mevalonic acid as an intermediate (Fig. 6.22).

The only antibiotic of any importance synthesized through the terpenoid pathway is fusidic acid, which has a steroid structure. It should be recalled that one product made by *Cephalosporium,* cephalosporin P, is a steroid and most likely has the same biogenetic origin. The most likely scheme for the biosynthesis of fusidic acid is the steroid pathway with squalene as the intermediate. This scheme is shown in Fig. 6.27.

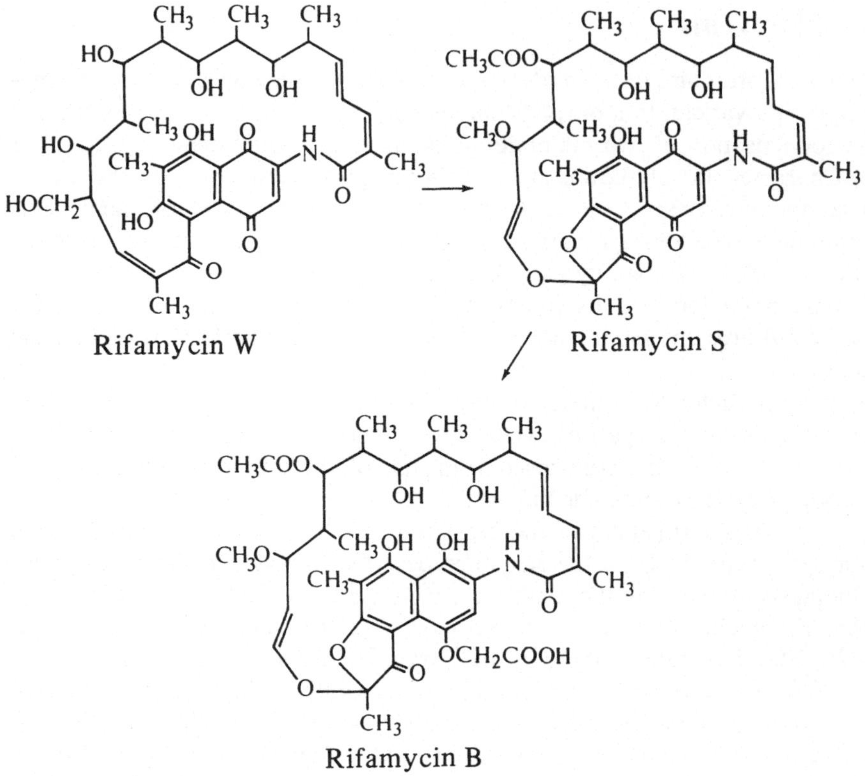

Rifamycin W

Rifamycin S

Rifamycin B

Figure 6.25. Biosynthesis of rifamycin B.

D. Aminoglycoside Antibiotics (Aminocyclitols)

Formally, the aminoglycoside antibiotics should not be called polymers but oligomers, since they consist of only a few units, usually three or four These units are a cyclic polyalcohol with six carbon atoms and some sugars, joined by glycoside linkages. Normally, some of the hydroxyls of the cyclitol and of the sugars are replaced by amine or substituted amine group, which gives rise to the class name. Several of the characteristics of this structure resemble those of some polysaccharides in the capsule and the bacterial cell wall, such as, for example, the lipopolysaccharides of the cell walls of the Enterobacteriaceae, with their long chains of amino sugars and amino alcohols.

One can therefore assume that these antibiotics are synthesized through the known pathways for the biosynthesis of the bacterial polysaccharides. These can be divided into two successive processes: (1) the series of re- actions that convert glucose into the individual units that make up the

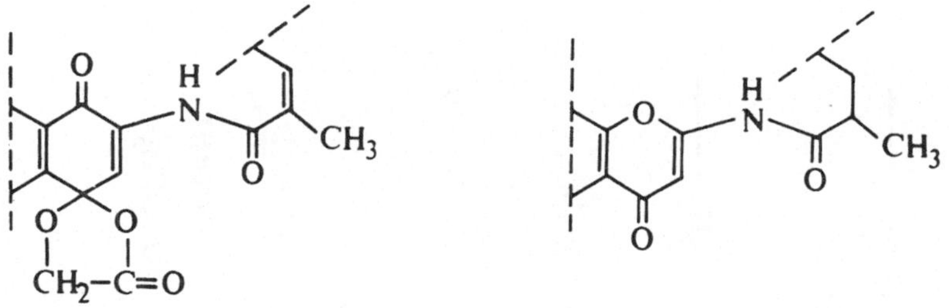

Figure 6.26. Rifamycin O and rifamycin G (partial structures, the rest of the molecule is the same as in rifamycin B).

chain and (2) the polymerization of these with formation of glycoside linkages.

Reactions in which one sugar is converted into another usually proceed without rearrangement of the order of the carbon atoms. In the cyclitols also, the order of the carbon atoms is the same as in glucose. Cyclitols are formed by formation of a bond between C-1 and C-6. In primary metabolism, sugars are then converted as sugar phosphatates or, more frequently, as sugars esterified in position 1 with a nucleotide (or deoxynucleotide) diphosphate, as in the classic example of the conversion of UDP-glucose into UDP-galactose. Esterification in position 1 with a nucleotide diphosphate is also the normal pathway of activation of a sugar, which must be done before it can form the glycoside linkage, and hence before oligomerization or polymerization.

We wish to emphasize that in all cases of biosynthesis of aminoglycoside antibiotics so far studied it has been possible to show that the individual units are made from glucose without rearrangement of the order of the carbon atoms. But it has not yet been possible to prove directly whether the glycoside bonds are formed after esterification with nucleotide diphosphates. However, there are some indirect indications of this and there is no reason to suggest the existence of a different mechanism.

The major aminoglycoside antibiotics can be divided into different groups according to the aminocyclitol (Table 5.1).

In spite of their structural similarities, the biosyntheses of the aminocyclitols are quite different. For example, deoxystreptamine is not a precursor of streptidine, and myoinositol, which is a precursor of streptidine, is not incorporated into deoxystreptamine. The biosynthesis of only a few aminoglycoside antibiotics has been studied, and of these, only that of streptomycin has been elucidated in any detail. We will summarize what is known about the biosynthesis of this antibiotic and indicate a few points about the biosynthesis of neomycin, a typical deoxystreptamine-containing antibiotic.

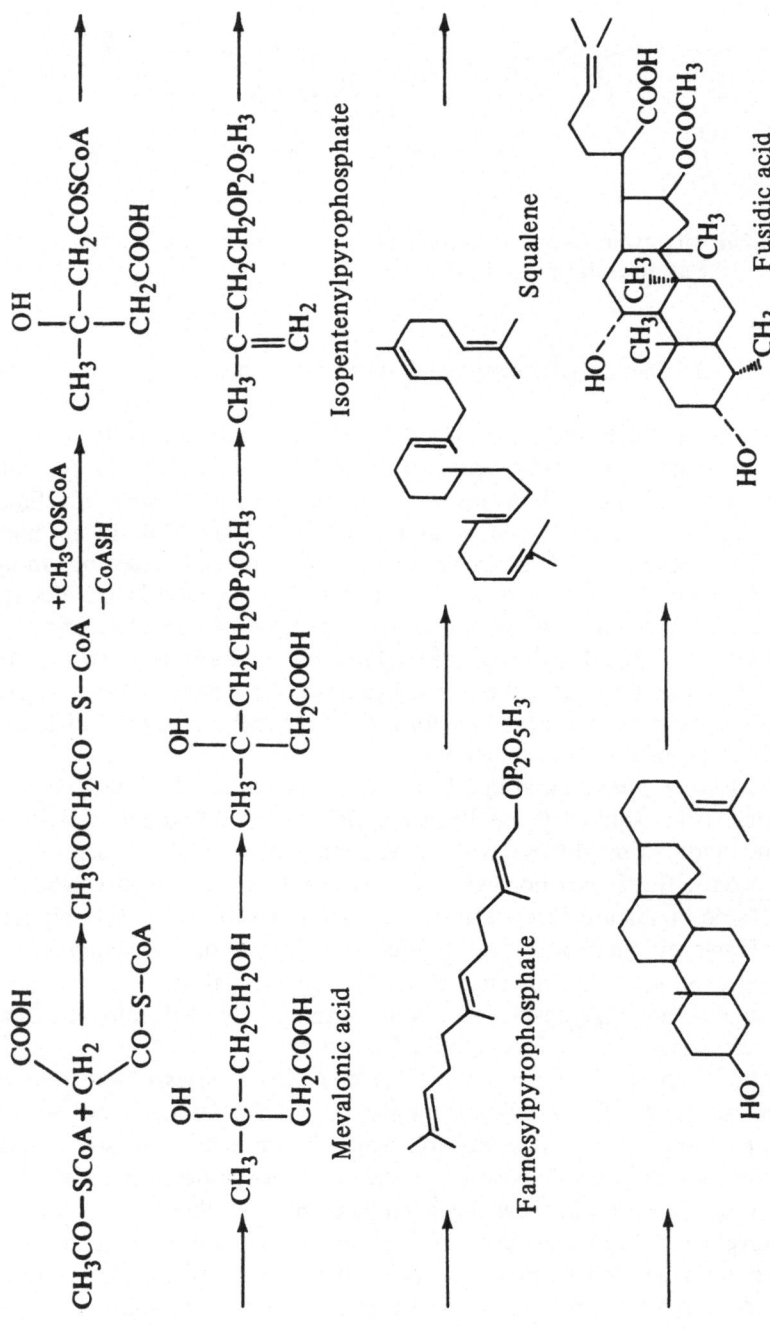

Figure 6.27. Probable biosynthesis of fusidic acid.

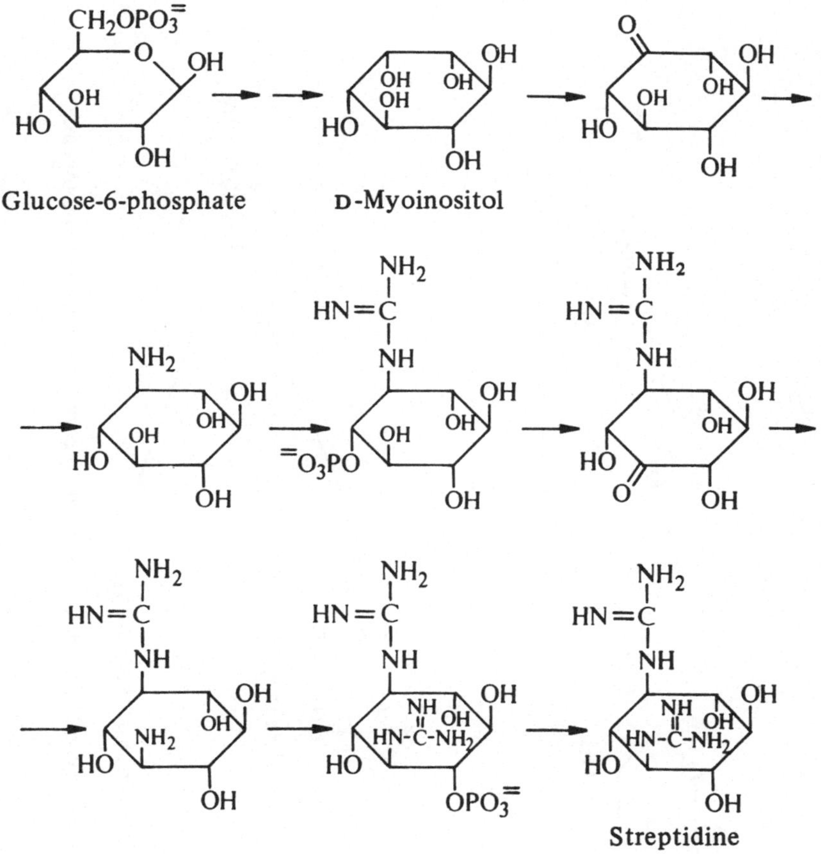

Figure 6.28. Biogenesis of streptidine.

1. Streptomycin

The molecule of streptomycin can be looked at as a trisaccharide made up of streptidine (an aminocyclitol), L-streptose and N-methyl-L-glucosamine. The biosynthesis of streptidine is the best studied. We know the reactions that lead to it from glucose (shown in somewhat simplified form in Fig. 6.28) and some of the enzymes that catalyze the reactions have been purified and characterized.

Note that neither streptamine nor deoxystreptamine is an intermediate in this biosynthetic pathway. The guanidine groups are introduced in two different series of reactions, each of which includes oxidation of a hydroxyl to a carbonyl, transamination and conversion of the amine group into a guanidine by reaction with arginine. This last reaction is preceded by a phosphorylation and followed by hydrolysis of the phosphate group.

We are not yet certain about the biosynthetic pathway of streptose.

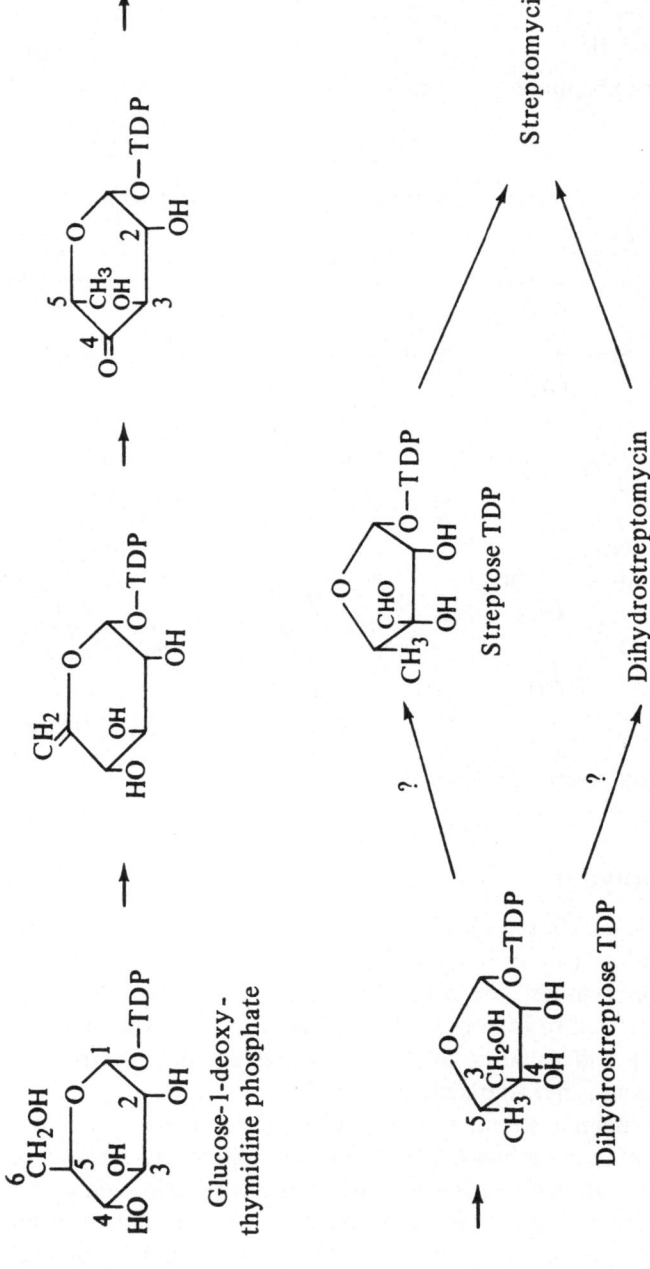

Figure 6.29. Biosynthesis of dihydrostreptose and hypothetical pathway of its incorporation into streptomycin.

However, we do know that streptose is made from glucose, with contraction of the ring by formation of a bond between C-2 and C-4, while C-3 becomes an aldehyde group. Recent studies indicate that this conversion takes place after esterification with deoxythymidine-5-diphosphate (dTDP) and that it is probably not streptose itself but its precursor dihydrostreptose that is incorporated into streptomycin (or, more exactly, into dihydrostreptomycin) and then later oxidized (Fig. 6.29).

N-methyl-L-glucosamine is made from glucose without rearrangement of the carbon atoms, and D-glucosamine (Fig. 6.30) is a probable intermediate. The reactions in which the configuration is inverted are not known. (Conversion of D-glucosamine into L-glucosamine requires epimerization of four asymmetric centers.)

Methylation of the amine takes place at the end, probably after the formation of the glycoside linkage to the other units (Fig. 6.31).

On the whole, it seems likely that the final steps in streptomycin biosynthesis are as shown in Fig. 6.31.

2. Neomycin

Although this is the best studied of the deoxystreptamine-containing antibiotics, only a few aspects of its biosynthesis are fully elucidated. As we have already said, deoxystreptamine differs in origin from streptidine. It

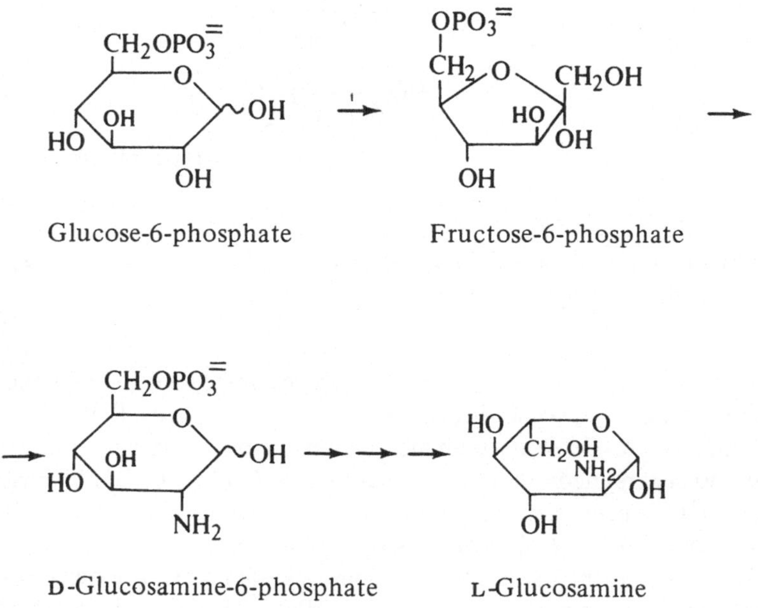

Figure 6.30. Origin of L-glucosamine.

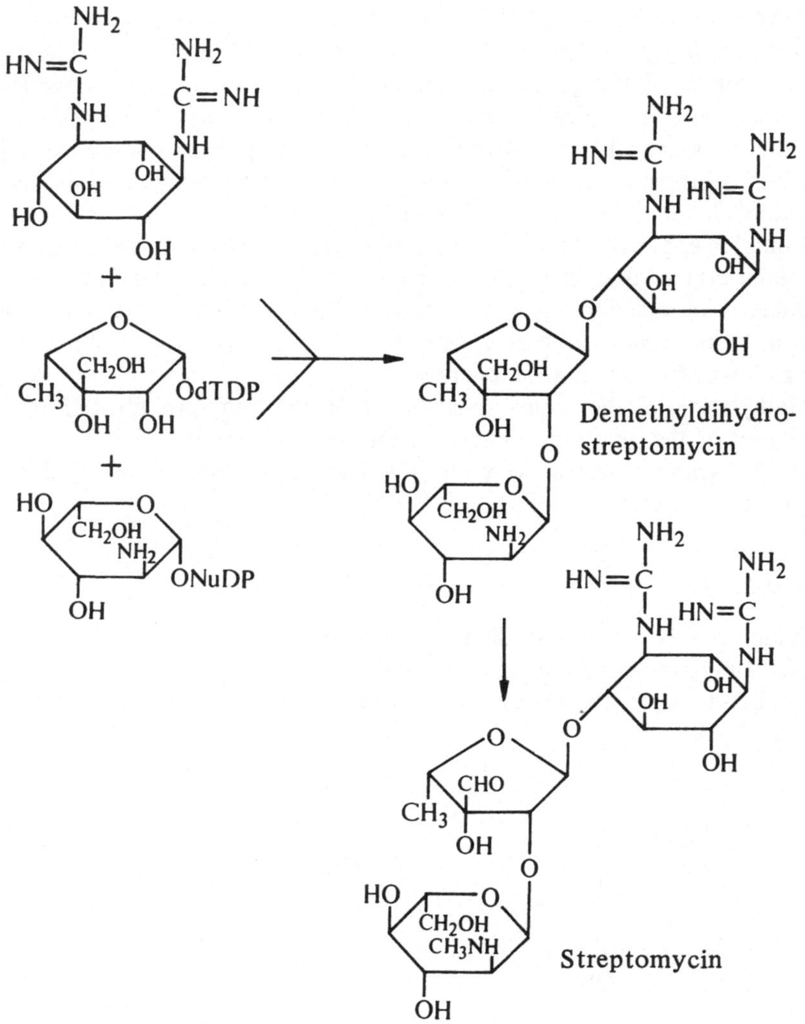

Figure 6.31. Biosynthesis of streptomycin. NuDP, nucleotide diphosphate; dTDP, deoxythymidine diphosphate.

was originally believed to originate from glucosamine, but today is thought to be derived from glucose through an inositol derivative (Fig. 6.32).

In addition to deoxystreptamine, neomycin also contains ribose and two amino sugars, neosamine B and neosamine C (Fig. 6.33). The ribose, as one would expect, comes from glucose, for the most part by elimination of C-1 through the pentosemonophosphate pathway.

The two amino sugars, neosamine B and neosamine C, are derived from glucose, without rearrangement of the carbons, through glucosamine. A hypothetical biosynthetic pathway is shown in Fig. 6.34.

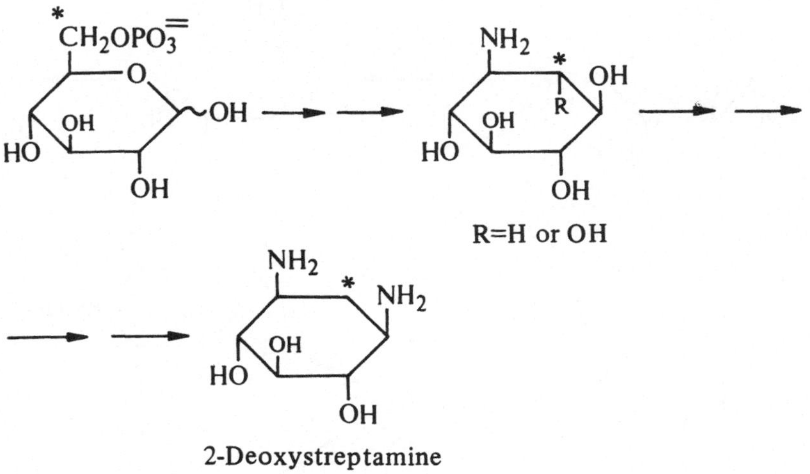

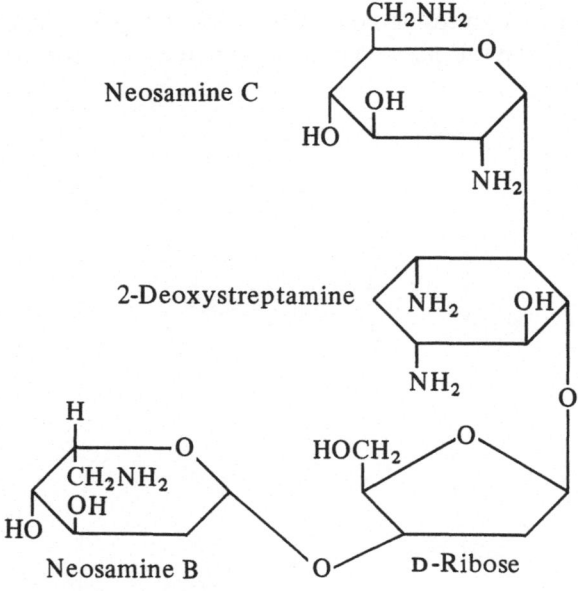

Figure 6.32. Hypothetical biosynthesis of 2-deoxystreptamine.

The arrangement of the rings to produce the final molecule is not known with certainty. However, isolation of di- and trisaccharides from mutants of *S. fradiae*, the neomycin-producing organism, and studies of incorporation of precursors seem to indicate the following sequence as

Figure 6.33. Structure of neomycin.

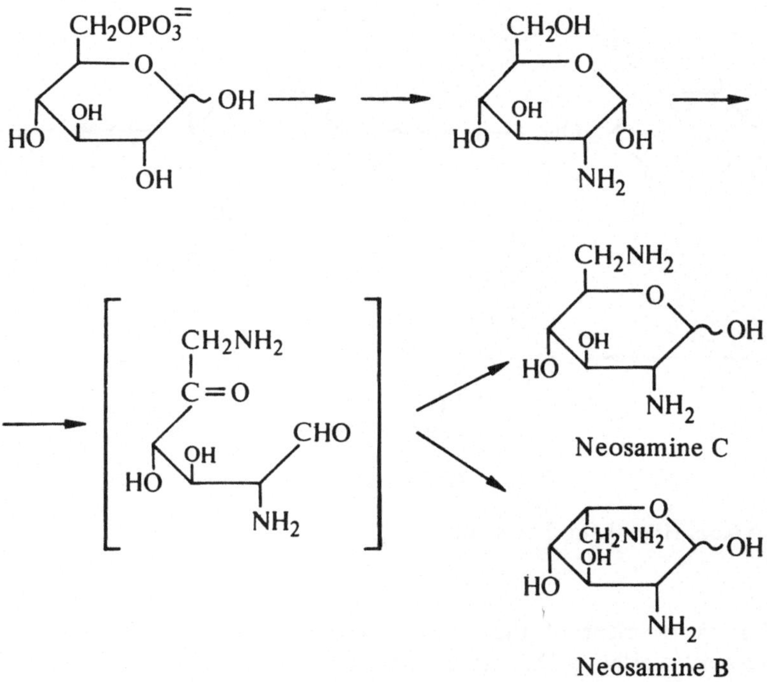

Figure 6.34. Hypothetical biosynthesis of neosamines.

probable: (1) Neosamine C is linked to deoxystreptamine to give neamine; (2) ribose forms a glycoside bond with neamine to give ribostamycin, (3) which is then linked to neosamine B to give the final structure.

Chapter 7

Search for and Development of New Antibiotics

I. Search for New Antibiotics

After the discovery and evaluation of penicillin, systematic programs to search for new antibiotics were initiated, and the results of these intensive screening activities are now well known.

How were these research programs carried out and how are they being carried out today? In this chapter we try to give brief answers to these questions.

A. Selection of Potential Antibiotic-Producing Strains

A *screening* program is one in which microbial strains that are potential antibiotic producers are selected and isolated as pure colonies. Experience has shown that by far the greatest number of antibiotic-producing microorganisms belong to species that live in the soil. For this reason, a large number of soil samples are obtained from all over the world and subjected to procedures for isolating the diverse microbial strains living in them. The usual scheme is as follows: 2–4 g of soil are well dispersed in aqueous suspension and drops of this suspension are placed on a nutrient agar medium and incubated until colonies of the microorganisms originally present in the earth sample appear. These are then isolated and cultivated in pure culture.

One then inoculates flasks of complex nutrient liquid medium with these and incubates the flasks for at least 2 or 3 days after growth has ceased. Thus synthesis and secretion of an antibiotic are facilitated for those strains that can produce it. The broths are then tested for antibiotic activity by one of the methods described in Chapter 2.

B. Extraction and Purification of the Antibiotic

Once a producing microbe has been found, it is necessary to extract and purify the substance responsible for the activity, so that all further tests for evaluation can be carried out with a sample as pure as possible. For this purpose, all the currently available techniques for chemical extraction of natural substances can be applied. Obviously, the procedures will vary for different antibiotics and will depend on the physicochemical properties, especially on the solubility characteristics in various solvents, of the particular substance.

C. Determination of the Novelty of the Antibiotic

Often in a research program to isolate new antibiotics, one detects a product already known. It is very important to establish at the start whether the antibiotic is new or already known. The problem is often very difficult, for two reasons:

1. The scientific literature contains descriptions of about 3000 different antibotics.
2. The producing microorganisms isolated in this way from the soil synthesize very small amounts of antibiotic that must be separated from large amounts of extraneous material in the culture broth.

An antibiotic is identified by systematic comparison of its properties with those of known antibiotics. This process is called dereplication. The principal properties of an antibiotic can be classified as follows (Table 7.1):

1. Microbiological properties, which include the antimicrobial spectrum, the spectrum of cross-resistance, and the spectrum of special effects, such as the effects of different test conditions (pH, presence of serum, concentration and kinds of ions, size of the inoculum, etc.) on the antibiotic activity; the frequency of development of resistant mutants together with the genetic nature of this (one step versus multistep); and the biochemical mechanism of the resistance give some additional indications of the nature of an unknown antibiotic.

2. Biological properties: The principal ones and those easiest to measure are the ED_{50} (see section II.A) and the LD_{50} (see section II.B) in the mouse, determined after various routes of administration. More important than the individual values is the ratio between the two, which is independ-

Table 7.1. Parameters for Identification of an Antibiotic

Microbiological Parameters
   Antibacterial spectrum
   Co-resistance spectrum
   Special effects spectra (pH, serum, ions, inoculum)
   Frequency of resistant mutants

Biological Parameters
   ED_{50}, different routes of administration, different pathogens
   LD_{50}, different routes of administration, different animals

Mechanism of Action
   Inhibition of synthesis of macromolecules in growing bacteria
   Inhibition of enzymatic reactions in cell-free systems

Solubility and Extractability
   Different solvents, pH effects
   Resins (anionic, cationic, weak, strong)

Chromatographic Behavior
   Paper, thin layer, with different solvent mixtures
   Electrophoresis
   Countercurrent distribution
   Liquid chromatography at high and low pressures

Physicochemical Properties
   Melting point
   Rotation of polarized light
   Elemental analysis
   Functional analysis
   Isoelectric point (electrofocusing)
   Spectra: IR, UV, visible, NMR, mass
   Stability: pH, light, specific enzymes

---

ent of the degree of purity of the antibiotic preparation being tested as long as the two parameters are determined with the same preparation.

3. Solubility properties in different solvents and the effect of pH on solubility. To these can also be added absorption onto different resins (anionic, cationic), which gives an indication of the molecular charge.

4. Chromatographic behavior, on different supports (paper, thin layer) and in different solvent mixtures. The antibiotic can be detected either by using specific reagents for certain functional groups (such as ninhydrin for antibiotics that contain the NH_2 group or Tollens' reagent for those containing reducing sugars), or by microbiological assay. The latter technique consists of overlaying a plate or paper strip chromatogram with an agar medium containing the test microorganism: After a suitable incubation time, the agar will have become turbid as a result of the bacterial growth except in the zones over the chromatogram in which the antibiotic was present and had diffused into the agar to inhibit bacterial growth. In

addition to chromatography, electrophoretic separation, which is usually faster, can also be used. Chromatography can be used not only for analysis but also for preparation of very small amounts of very pure product.

5. Physicochemical properties such as melting point, rotation of polarized light, elemental analysis, analysis of principal functional groups, infrared spectrum, ultraviolet spectrum, visible light spectrum, NMR spectrum, and mass spectrum. Another important property for identification is the stability of the antibiotic activity after exposure to different factors such as light, pH, and specific enzymes.

6. Mechanism of action. The determination of even an approximation of the mechanism of action has essentially two functions: an identification function, which is of great help to the chemist, as it enables him to focus his program on comparison with known products with that mechanism of action, and it gives some clues about the intrinsic toxicity of the test product.

D. Originality and Probability of Success

The screening procedures described do not differ significantly with the exception of some technical innovations, from those carried out in the initial screening programs some 30 years ago. But it has become ever more difficult to develop and introduce new antibiotics into medical practice for two major reasons. First, a new antibiotic can be placed on the market only if it has properties that are superior to those of already existing products. Second, the criteria for admission of new products set up by the various health authorities that regulate this in different countries have become ever more stringent, excluding from the market many new products that would have been accepted in previous years. As a consequence, the probability of a successful outcome from a screening program becomes smaller and smaller. For this reason, there is also a search for technical or conceptual innovations that might be introduced into a screening program to increase the probability of success. The proposals advanced by the various experts deal with three categories of factors that can be varied: (1) quantitative factors, (2) qualitative factors, and (3) organizational factors.

1. Quantitative Factors

The influence of quantitative factors is straightforward. All other factors being equal, the more fermentation broths studied, the greater is the probability of finding a useful substance. The quantitative dimensions of a program are limited by obvious considerations of cost. However, the number of strains isolated and studied might well be increased considerably by suitable application of automated techniques and microtesting procedures (see Chapter 2).

2. Qualitative Factors

These can be subdivided into three classes: (1) Screening novel organisms, (2) varying the culture conditions, and (3) using new selective test procedures.

a. Screening Novel Organisms

The opinion is widely held that the probability of discovering a new antibiotic is very small if we continue to isolate and test the same microorganisms that have been screened over the last 30 years.

This opinion is based on the frustrating, unfortunately frequent, experience of rediscovering an already known antibiotic. At least 7 million organisms have been isolated, more than 3000 different antibiotics have been described and partially characterized, and about 100 new ones are described each year. However, many experts are convinced that not all the types of antibiotic that exist in nature have been discovered, and that the chance of finding new chemical structures would be much greater if we would examine unusual, rare microorganisms. This opinion is based on practical experience and on the idea that the structures of microbial secondary metabolites are expressions of the genetic characteristics of the producing species.

Two problems arise from this proposition, one theoretical and one practical. The theoretical one may be stated in these terms: How far apart on the taxonomic scale, or perhaps, more exactly, on the evolutionary scale, must two groups of microorganisms be before their secondary metabolites will be sufficiently different? It is already known that secondary metabolites, including those with antibiotic activity, are quite different in the bacilli, actinomycetes, and lower fungi (the three large groups of antibiotic producers). All three of these groups have been subjected to extensive screening. Therefore, it is necessary to select within these three groups subgroups (genera) that have not been studied in detail. This leads to the practical problem. The rare genera have not been examined in detail because they are difficult to isolate and/or grow.

On different occasions many techniques and selection programs have been described that presumably would serve to select from soil rare genera, especially of actinomycetes.

b. Varying the Culture

In every screening program, the first step is isolation of microorganisms from the soil followed by growth, usually in liquid medium, of the colonies to be tested. Let us assume that the colonies isolated belong to a rare genus that presumably differs genetically from the common ones. It is known that secondary metabolites are not indispensable substances and need not be produced by a microorganism because it has the genetic information to do so. Many secondary metabolites are produced during the idiophase of the culture, i.e., when growth has ended. In many cases their synthesis is controlled by a mechanism such as final product repression, C-catabolite repression, N-repression, energy-charge, or PO_4 repression. It is not sufficient to have microorganisms available with a novel genome, but it will be essential to grow them under conditions that permit maximal expression of the genome.

In general, these conditions are not those that support rapid growth. Therefore, one must incubate the microorganisms for some time after growth has stopped and make certain that in the idiophase the levels of carbon and nitrogen sources and the PO_4 concentration are not too high.

c. Use of New Test Procedures

Other things being equal, the probability of finding a new antibiotic will depend on the test used for demonstrating it in the sense that the probability is greater if one uses a test that reveals activities that are "selected against" in the more commonly used tests.

As examples, it is well known that synthetic test media reveal antimetabolites that are often inactive in complex media. The pH of the test medium tends to select substances with isoelectric points near the pH of the medium, as antibiotics often have maximal activity at the isoelectric point.

3. Organizational Factors

These are factors that determine how a research program should be established. A successful antibiotic must have the following minimal characteristics: (1) It must be *new,* and *patentable;* (2) it must *cure* infections in animals; and (3) it must not have any *serious toxic effects* at the doses used for treatment.

Although the programs described above should increase the probability of finding a new active product, they obviously cannot guarantee that every activity found will be new. Nor will they assure in any way that products active in vitro as inhibitors of bacterial growth will also be effective in cure of experimental infections or that they will be nontoxic. These three fundamental characteristics must be demonstrated directly for any presumptive new product.

We can present the problem as it stands at this stage by asking other questions. Should one first determine the effectiveness and the nontoxicity of product, followed by determination of its novelty, or should one do this in the reverse order? Also, since the properties that characterize an antibiotic and distinguish it from all the others are numerous and must be obtained by different types of methodology that give different types of information, which ones should be studied first and which later? These two questions serve to illustrate that a primary screening program must be comprised of a group of subprograms using different techniques, aimed at obtaining information about one or more of the three essential properties (novelty, effectiveness, nontoxicity). It is important to decide on the organization of these programs that will be most efficient, i.e., will give the desired information with the least amount of effort, and with the lowest cost.

Screening programs in different research laboratories will differ in sequence, primarily because of differences in personal preferences or the skills available, and on the experience of each research group. However, some general observations can be made: (1) Determination of the novelty of the product is usually more complex than determination of its effectiveness or of its nontoxicity; (2) the various parameters that characterize an antibiotic have different values for identification, in addition to the fact that some can be determined for relatively crude products and others require a high degree of purity.

The cost of a test is strictly correlated with the quantity, and even more with the degree of purity, of the sample analyzed. An ideal way to conduct a screening program would be to determine the characteristics of the product, together with a reliable identification, using a small amount of crude product.

E. Screening of Congeners of Known Products

The screening program we have been referring to has been aimed at discovery of new structures with antibiotic activity. A screening program may also have as its objective the isolation of compounds similar to known antibiotics, but with small structural modifications that might improve the properties. Such substances are called *congeners* of the known compounds.

This objective can be approached in three ways: (1) by synthesis, (2) by mutation, (3) by fermentation.

1. Synthetic Approach

This consists of chemical transformation of a known antibiotic. The basic concepts and the objectives of this approach are described in Chapter 5.

2. Mutational Approach

As described in Chapter 6, antibiotics, like other secondary metabolites, tend to be produced in families of metabolically related substances. Introduction of mutations into the wild strain may favor the biosynthesis of products that are present in trace amounts or not present at all in the original strain.

3. Biosynthetic Approach

Some producing strains are able to incorporate into the antibiotic molecules precursors that are similar to the "natural" precursors synthesized
by the strain itself. A good example is that of the synthesis of penicillin V
after addition of phenylacetate to the culture broth of the strain that produces penicillin G. One situation that combines the mutational approach
with the fermentation approach is so-called *mutational biosynthesis* or
mutasynthesis. This consists of the genetic conversion of a producing
strain into a mutant, termed an *idiotroph,* that has lost the ability to synthesize the antibiotic unless some constituent of the molecule is added to
the culture medium. By providing this strain with analogs of the essential
constituent one can obtain different antibiotics. This last approach has
been used for some years in the screening for new aminoglycoside antibiotics.

II. Development of an Antibiotic from the Laboratory to the Clinic

Let us imagine that the screening program whose prime focus is the discovery of new antibiotics, described in the preceding section, has produced
one or more substances with interesting antibacterial activity in the in vitro
tests and that we want to develop it or them into products to be used in
medical practice. Obviously, one cannot go directly into clinical trials with
new products active in vitro for ethical reasons: The only products that
should be given experimentally to man are those that have a high probability of being useful and whose tolerability has been ascertained in different animal species.

Therefore, possible candidates must be selected among the available
compounds by determination of their effectiveness in curing animals infected with pathogenic bacteria; i.e., the antibiotic must cure experimental
infections. As is always the case in new drug research, the experimental
model never correlates with the real clinical disease. Therefore, the information obtained from experimental infections does not have absolute
predictive value, but it does increase the chance of leading one to make a
better decision, especially when one correlates toxicity data for the drug in
the same animal species.

A. Experimental Infections

The experimental infection is produced by injecting into an animal, under
standard conditions, a known number of pathogenic bacteria of known
virulence. Usually, the ideal laboratory host is the mouse, which costs

little and in which the reactions to infection are well established. There are different models of experimental infection. The one most commonly used is to produce a septicemia (bacteria diffused through the circulation into many organs and tissues) by injecting the mouse intraperitoneally with a culture of pathogenic bacteria, and by determining the daily dose of antibiotic that, when given by the route being studied, will cure 50% of the treated animals. This *effective dose* is called the ED_{50} or the CD_{50} (effective dose 50 or curative dose 50) and is expressed in mg/kg, i.e., in milligrams of antibiotic per kilogram of body weight of the animal. The number of infective bacteria must be sufficiently large to kill all the untreated infected animals within a given time, usually 2–3 days. Different types of information about the pharmacological properties of the antibiotic can be obtained by varying the route of administration and the time interval between infection and treatment. The most common schemes for treatment of experimental septicemia are the following:

1. I.P.–S.C. (I.M.): This consists of administration of the pathogenic agent intraperitoneally (I.P.) and the antibiotic subcutaneously (S.C.) or intramuscularly (I.M.). If the product is active, one can conclude that: (1) it has the capacity to travel from the injection site to the different infected tissues (the product is absorbed from under the skin or the deposit in the muscle); (2) it is present in the blood and in the tissues in an active form at high enough a concentration for long enough a time to inhibit the growth of the bacteria, before it is metabolized and excreted.

2. I.P.–P.O.: This consists of giving the pathogen intraperitoneally and the antibiotic orally (P.O.). A cure demonstrates the capacity of the product to be absorbed from the stomach or the intestine and to be diffused through the blood into the tissues in concentrations sufficient to inhibit bacterial growth. In addition, it indicates that the product is not inactivated by the acidity or the hydrolytic enzymes present in the stomach.

Occasionally an I.P.–I.P. scheme is used, with both pathogen and antibiotic given intraperitoneally. When this treatment protects the mice from the infection, one concludes that the antibiotic is effective against the infecting organisms in the presence of living cells and enzymes; i.e., it is not inactivated by animal cells. Aside from this, the I.P.–I.P. test is not very different from a test in a test tube, with the peritoneum of the mouse serving as the test tube. Therefore, it has very little predictive value for the true effectiveness of the antibiotic.

In any of these schemes, the treatment may be given at the same time as or immediately after administration of the pathogen or after various intervals. In the first case, the prophylactic power is determined and in the second the true curative capacity of the product.

In the evaluation of the results of the experimental infection test, it must be kept in mind that its *predictive value is essentially qualitative,* not quantitative. It indicates the probability that an antibiotic will be effective but gives no definite indication about the dosage that will be sufficient in man.

The septicemia model has the advantage that it affords reproducible results in different laboratories, and is, therefore, a useful method for comparing different antibiotics, especially if they are structurally similar.

However, this model does not serve for certain types of infection, especially those that are chronic and localized, such as urinary tract infections or pulmonary tuberculosis. For those specific applications, special models must be used, such as pyelonephritis (infection of the urinary tract and kidneys) in the rat, which is obtained by different techniques and which has some of the properties of a chronic infection, or experimental tuberculosis of the guinea pig or mouse, also slow-developing. In these cases, the effective dose is not determined solely on the basis of the percentage of surviving animals. It also takes into consideration other parameters such as the bacterial count at the site of infection and the duration of survival.

Another type of experimental infection often used in the laboratory is the so-called "topical infection," which includes infections of skin wounds produced surgically or by burning and infections of the cornea of the rabbit's eye.

The antibiotic can be given parenterally or applied topically to the infection site. Evaluation of the effectiveness will be different for different types of infection. When the infection causes death of untreated controls, the effectiveness is expressed as the ED_{50}. When it does not, the criterion for cure is based on the differences in rates of healing of the wound in treated and untreated infected animals.

B. In Vitro Activity versus In Vivo Activity

There are now 3000 antibiotics described, but of which not more than 10%–15% are active in curing experimental infections. The most important reasons for in vivo inactivity of substances active in vitro are listed in Table 7.2.

C. Toxicity

For an antibiotic to be introduced into medical practice, it must be relatively nontoxic to the patient. This lack of toxicity is relative, since any product may be toxic if given in excessive doses. It is important that there be a reasonable margin of safety between the doses that are effective in treatment and those at which toxic signs appear.

For obvious reasons, the potential toxicity of a drug in man must be extrapolated from experiments done in animals, which poses the problem of specific toxic effects in some animal species, briefly discussed later.

Some indication of the toxicity of a drug is obtained by determining the

Table 7.2. Reasons Why Substances Active in Vitro May Be Inactive in Vivo

Microbiological	Metabolic state of bacteria (dormant state, anaerobic condition, etc.).
	Presence of nonpathogenic microbes that can inactivate the antibiotic.
Pharmacokinetic	The antibiotic is poorly absorbed or rapidly excreted.
	The antibiotic is rapidly metabolized to inactive products.
	The antibiotic cannot reach the bacteria located in abscesses, fibrin deposits, intracellularly, in bones, etc.
Toxicity	Doses that could be effective cannot be given because they are toxic to the host.
Biochemical	The biochemical environment of the bacteria may antagonize the antibiotic (pH, antagonism by antimetabolites).
	Strong binding by serum proteins is a common cause of in vivo inactivity.

so-called *acute toxicity,* which is usually expressed as the LD_{50}, the dose in mg/kg of body weight which is lethal for 50% of the animals treated. The LD_{50} is determined after administration of single doses of different concentrations to groups of animals (usually mice). From the percentages of deaths observed at the different doses, one extrapolates to the dose that would have caused the deaths of 50% of the animals. The LD_{50} varies with the route of administration (oral, subcutaneous, intramuscular, intraperitoneal, intravenous), and this must always be indicated along with the species of animal used. Table 7.3 shows the acute LD_{50} values in the mouse for some antibiotics.

A more in-depth evaluation can be made from the study of subacute

Table 7.3. Values of LD_{50} and ED_{50} for Some Antibiotics Given by Different Routes to the Mouse[a]

Antibiotic	MIC (μg/ml)	LD$_{50}$ OS	LD$_{50}$ I.P.	LD$_{50}$ I.V.	ED$_{50}$ OS	ED$_{50}$ S.C.
Chloramphenicol	4	2460	1320	100–200	116	26
Penicillin G	0.05	>5000	3490	2340	1.7	0.1
Erythromycin	0.1	2927	660	426	48	5.8
Streptomycin	10	>5000	1400	120	80	1.2
Lincomycin	0.5	—	1000	214	40	30
Tetracycline	0.5	3000	200–300	160	12	6
Rifampin	0.005	770	340	585	0.12	0.11
Amphotericin B	n.a.	—	280	7	n.a.	n.a.
Ampicillin	0.1	>5000	3400	—	3	11

[a] The experimental infection was *S. aureus* septicemia.

and chronic toxicity, which involve daily administration of different doses of the product for either 1–3 months or for 6–24 months. These experiments are carried out in at least two species of animals, of which one must be a nonrodent mammal. The aim of this type of study is to (1) determine the maximal safe dose, which results in no toxic effects, (2) determine which organs and functions are most damaged by large doses of the product.

One also attempts to discover the less obvious toxic manifestations by observing carefully during the entire treatment period the behavior of the treated animals, the consumption of food and water, body growth, and the appearance of macroscopically visible pathological changes. At specific intervals, blood chemistry determinations and various organ function tests are performed. At the end of the treatment, the animals are sacrificed and the individual organs examined both macroscopically and histologically.

It is possible to use data obtained in one species of animal to predict the toxicity of the product in another species, particularly in man, because of the biological fact that there are considerable similarities between species with regard to the basic biochemical processes of cellular metabolism.

There are certainly no substantial differences between species in protein synthesis, DNA replication, glycolysis, oxidative phosphorylation, etc. However, some cases have been found in which the toxicity of a product is quite different in different species. With rare exceptions (because it is impossible to foresee these differences, the initial administration of a drug to man is done very cautiously), the causes of these variations in toxicity, which can usually be seen more clearly during chronic treatment, lie in differing capacities of the different species to absorb, excrete, and metabolize the exogenous substances. These capacities have evolved differently in different mammalian species, usually in relation to the different dietary requirements.

Therefore, it is not surprising that an exogenous substance such as an antibiotic is absorbed or excreted at different rates in different species, and that in some cases, especially when the antibiotic is administered repeatedly and in large doses, it accumulates to a point at which it has damaging effects. For this reason, toxicological studies must be accompanied by comparative pharmacokinetic studies.

In the extrapolation of toxicity data from laboratory animals to man, one must remember that there are differences in the normal bacterial flora in different species. A classic example of the importance of this is the high toxicity of repeated doses of penicillin in guinea pigs, although it is one of the least toxic antibiotics in man. Penicillin kills the Gram-positive flora in the guinea pig intestine, which is then colonized by Gram-negative bacteria, which results in a fatal bacteremia. The intestinal flora of man is quite different from that of the guinea pig, and if the toxicity of penicillin had been tested only in the guinea pig, this product would probably have been discarded.

D. Pharmacokinetics

Pharmacokinetics is the study of the absorption, distribution, and excretion of a drug in a living animal. It also includes the study of the metabolism of the drug, i.e., the chemical transformations it undergoes in the body, usually through enzymatic reactions.

Pharmacokinetics is extremely important for determining the possible therapeutic application of an antibiotic for two reasons:

1. To be effective, an antibiotic must reach the infection site in concentrations greater than the MIC and remain there for a sufficiently long time. Study of the pharmacokinetics tells us the concentrations in the various tissues and organs as a function of time, permits prediction of the therapeutic effectiveness, and gives initial indication of the doses that should be used and how they should be given in later clinical studies.
2. Comparing the pharmacokinetics in different animals with those in man helps to determine in which species the toxicity studies should be most applicable to man.

We must emphasize that from this point of view antibiotics are not a homogeneous family. For every antibiotic one must determine which of the species most commonly used has pharmacokinetic and metabolic properties most like those of man.

The study of pharmacokinetics of antibiotics is easier than for other drugs because their antimicrobial activities can be used to determine the concentrations of the product in biological fluids and tissues. However, a complete pharmacokinetic and metabolic study must include one with a radioactive antibiotic, which is usually obtained by fermentation whenever there is available a strain that is a good producer and one has an idea of the biosynthetic pathway and the principal precursors of the antibiotic.

E. Clinical Studies

Clinical trials establish the therapeutic usefulness of an antibiotic, the adverse reactions it causes, and the optimal dose at which it should be given.

A clinical trial begins with the administration of single doses to healthy volunteers to assure that the product is well tolerated at high blood levels. These data are essential for determining the dose to use in the initial therapeutic trials. We must keep in mind that the CD_{50}, the curative dose determined in experimentally infected animals, is not a good indication of the dose that will be effective in man, given the differences between a fulminating experimental septicemia and the slower course of normal clinical infections and the differences in pharmacokinetics. Usually, in experimental infections the bacteriostatic antibiotics appear to be active at higher doses and the bactericidal ones at lower doses than those needed in patients. Therefore, the initial dose and schedule for the therapeutic trials are established to obtain relatively constant blood levels greater than the MIC during the entire treatment, especially when the antibiotic is bacteriostatic. For the bactericidal antibiotics it is more important to obtain high blood levels than a constant blood level.

Because of the numerous objective symptoms in infections (decrease in fever, disappearance of bacteria, etc.), the initial trials can be carried out under so-called *open* conditions, in which the physician can judge the results obtained with the product and compare them with those he has seen with other antibiotics. Later, the validity of the antibiotic must be determined exactly in *controlled* trials, in which the effectiveness is compared with that of another antibiotic already in clinical use without the physician having knowledge of which of the two products is being given at any given time, so that there will be no effect of this knowledge on his judgment.

Unlike for other drugs, in controlled trials of antibiotics one does not include the third group, those treated with placebo (by definition a substance innocuous and ineffective), for obvious ethical reasons.

Another aim of the clinical trial is to evaluate the tolerability and the adverse reactions. For these purposes, the controlled experiments yield more reliable data, because the adverse reactions are often evaluated from subjective criteria and may be due to causes other than the treatment, i.e., to the infectious disease itself or to a concomitant disease.

Table 7.4 outlines the schedule for the different types of experiment and the usual time needed from the initial finding to the introduction of the antibiotic into medical practice. The time and the costs of these developments tend to increase continuously, among other reasons because of the ever greater requirements of the health authorities of different countries. The clinical trials and the toxicological studies must be carried out according to protocols established by the health authorities of different countries.

III. Development: From the Laboratory to the Manufacturing Process

Antibiotic-producing strains originally isolated from soil usually synthesize less than 50 mg of product per liter of culture. To carry out all the studies necessary for evaluating the compound, several kilograms are needed.

Moreover, if the product is introduced successfully into clinical practice, tons of it must be produced every year. The cost of the product would be prohibitive were the productivity limited to that of the original strain. Obviously, one of the first steps in development of the antibiotic must be to establish means to increase biosynthetic productivity. These studies may

Table 7.4. Simplified Scheme of Antibiotic Development

Development of the Producing Strain	Discovery and Development of the Antibiotic	
Wild-type strain	Microorganisms isolated from soil ↓ Fermentation ↓ Activity assay ↓	Year zero
Studies on culture media and conditions	Extraction of the active product ↓ Evaluation of: Novelty, efficacy in experimental infection, acute toxicity, absence of mutagenic properties, local tolerability ↓ Patent	Year 2
Mutagenesis and selection of high-producing strains	3-Months subacute toxicity on two animal species Clinical tolerability trials on healthy volunteers → Efficacy trials on humans (dose finding) Pharmacokinetics and metabolism → Chronic toxicity on two species	
Pilot plant fermentation studies	CLINICAL EXPERIMENTATION ON HUMANS (ABOUT 1000 CASES)	Year 5
Industrial production	EFFICACY ←→ SIDE-EFFECTS MONITORING ↓ APPLICATION FOR REGISTRATION	Year 7–8

be approached in two different ways: to improve the production strain and to improve the conditions of fermentation.

A. Improvement of the Producing Strain

By this term (or the often used expression, development of the producing strain) is meant manipulations of the strain itself to increase its capacity to produce antibiotic. There are two types of manipulation: (1) mutation and selection and (2) genetic engineering.

1. Mutation and Selection

Production of secondary metabolites, which is what antibiotics are, is controlled by more than one mechanism. The rate of production of such metabolites depends on the rates of synthesis of the metabolic products that are intermediates in the antibiotic synthesis and on the rates of their transformation. These reactions are regulated by mechanisms such as repression or induction of enzyme synthesis, allosteric regulation of the activity of the enzymes, or catabolite repression, each of which is determined genetically. Therefore, these mechanisms can be changed or even eliminated by mutation (a stable change in the DNA structure). Mutations occur naturally in a bacterium during cell division, but at a very low frequency. The frequency of mutation can be increased by several orders of magnitude by subjecting the strain to mutagenic agents such as ultraviolet radiation, X-rays, or chemical agents (nitrous acid, nitrosoguanidine, etc.).

Most mutations cause changes in the DNA that are lethal to the cell, but some that cause changes in certain sectors of its metabolism still allow the cell (called a mutant) to live and may also cause an increase in antibiotic production. The population of microorganisms surviving after mutagenic treatment (i.e., those cells that have not undergone lethal mutations) consist of: (1) cells in which there has been no mutation and that are therefore like the original strain; (2) cells in which there have been produced one or more mutations that either do not interfere with antibiotic synthesis or that inhibit antibiotic synthesis; (3) cells that have undergone mutations that increase the production of antibiotic. These last cells, called high-production mutants, are present in the population with very low frequency. The problem consists of identifying these high-producers in a mass of low-producers. The process by which this is done is called *selection*.

a. Random Selection

This is the most empirical method of selection. Suspensions of organisms that have been subjected to mutagenic treatment are plated out to give colonies derived from single cells (uniclonal colonies). Certain ones of these colonies are selected blindly for cultivation in pure culture and then the potency (the concentration of an antibiotic produced in a fixed time of fermentation) is determined. Since mutation toward high productivity is a rare event, a large number of colonies must be tested to find the high-producing mutant. A program of this type requires selection and testing of about 10,000 colonies per year. Even though this is not a very efficient or intellectually satisfying procedure, it is often the only feasible procedure, at least during the early phases of development.

The efficiency of this type of selection can be increased by preselecting the potential high-producers directly on the plate. This can be done by covering the surface of the plate, after the colonies have grown, with a layer of agar containing a bacterium sensitive to the antibiotic. The colonies of high-producing organisms show larger inhibition haloes than those around the colonies of the original strain.

b. Selection of Morphological Mutants

Among the survivors of mutagenic treatment some colonies appear that are morphologically different from the wild type. In some cases a qualitative relationship has been found between the morphology of the colony and overproduction of antibiotic (e.g., penicillin, cycloheximide, nystatin). However, there has never been any explanation for the fact that a given morphological mutant produces more antibiotic. It is possible that the effect of the mutation that leads to morphological change was located in one of the enzymes for sugar metabolism, with a consequent change in cellular levels of phosphorylated sugars, of coenzymes, and of the ratio ATP/ADP. Changes in these could cause, on the one hand, a change in the composition of the cell wall, and thus a change in the morphology of the colony and, on the other hand, a change in the rate of synthesis of precursors for the antibiotic.

c. Selection of Mutants That Are High-Producers of Intermediate Metabolites

Secondary metabolites are derived by transformation or polymerization of intermediate metabolites. An increased rate of synthesis of these latter might lead to greater productivity. For example, the rate of methylation of an antibiotic might be limited by the availability of the methyl donor, which is often the amino acid methionine. If the pool of methionine is increased by a mutation which leads to overproduction of this amino acid, it would lead to an increased production of the antibiotic.

d. Selection of Auxotrophic Mutants and Their Revertants

An auxotroph is a mutant that differs from the original strain in requiring an additional nutrient. For example, an auxotroph for vitamin B_{12} must be given this substance in order to grow. It has been seen that in many cases auxotrophy has a profound effect on antibiotic synthesis. In general, auxotrophic mutants produce less than the wild strain. However, if the required substance is added to the culture medium of some auxotrophic mutants, a notable stimulation of production to levels greater than those in the wild strain has been found. This same result can on occasion be obtained by producing first the auxotrophic mutant and then the revertant mutant, i.e., a strain that no longer requires addition of the nutrient for growth. Even though we do not know for any individual case the mechanism by which reversion from auxotrophy to the primary condition (prototrophy) causes an increase in antibiotic production, we can suppose that this could be due to a change in the control mechanisms of one or more metabolic pathways, with consequently greater production of one or more precursors of the antibiotic, and thus with stimulation of the production of the antibiotic.

In conclusion:

1. Auxotrophy usually leads to mutants which are low-producers.
2. On occasion, auxotrophic mutants produce more than the original strain if the medium is supplemented with the required nutrients.
3. Some revertants produce more than the original strain.

e. Selection of Mutants with Changes in Their Metabolic Control Mechanisms

The synthesis of primary and secondary metabolites by a microorganism is regulated by many control mechanisms that act to prevent excessive production. Since the aim of the project for improvement of the productive strain consists exactly in changing it so as to obtain large amounts of a given metabolite, it is easy to understand how elimination of regulatory mechanisms might be important for attaining this end. The principal control mechanisms are the following:

1. *End-product feedback*. This is the inhibition of the *activity* (not the synthesis) of an enzyme (usually the first of a metabolic pathway) brought about by a small molecule, called the effector. Such an effector could be (see Fig. 7.1):

a. The final product of the metabolic path, i.e., the antibiotic itself as in the case of chloramphenicol which inhibits the activity of the enzyme arylamine synthetase. This is the first enzyme of a pathway branching off the common pathway of aromatic amino acid synthesis (end-product inhibition).
b. A precursor of the antibiotic. A precursor of the antibiotic may feedback-regulate its own synthesis and therefore limit the rate of synthesis of the antibiotic.
c. A metabolite sharing part of the antibiotic synthesizing pathway.

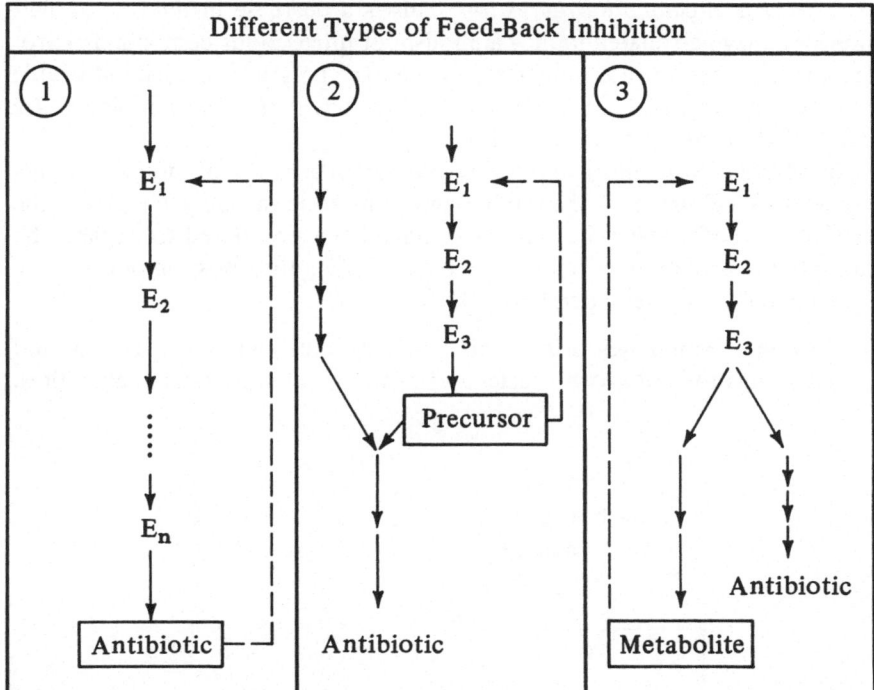

Figure 7.1. Different types of feedback inhibition.

2. *Repression of synthesis of one or more enzymes.* This is the inhibition of the rate of synthesis of one or more enzymes of the synthetic pathway of the antibiotic. The effector is a small molecule, usually the final product of the enzymatic reactions or a molecule structurally similar to this product.

3. *Induction of the synthesis of one or more enzymes.* This is an increase in the rate of synthesis of one or more enzymes caused specifically by a small molecule that is generally the substrate of the enzyme or a structurally similar molecule. For example, lactose induces the synthesis of β-galactosidase, which splits lactose into its component sugars, glucose and galactose.

4. *Glucose-catabolite repression.* This is a decrease in the rate of synthesis of some enzymes, especially those of degradative metabolism (glycolytic enzymes of the pentose phosphate cycle, enzymes of the Entner-Doudoroff pathway, of the glyoxylate cycle, of amino acid degradation) in the presence of glucose or other sources of easily utilizable carbon.

As a consequence of the repression of the synthesis of these enzymes, there is a decrease in the synthesis of metabolites in the pathways that include these enzymes. Synthesis of antibiotics is often under catabolite repression control.

5. *Nitrogen-catabolite repression.* This is a decrease in the rate of synthesis of enzymes related to the catabolism of nitrogenous compounds (proteases, amidases, nitrate reductase, etc.) in the presence of easily utilizable nitrogen sources, such as ammonia. The syntheses of many antibiotics are controlled by this type of mechanism.

6. *Regulation by inorganic phosphate.* Often addition of inorganic phosphate to a culture actively synthesizing antibiotic strongly decreases the rate of synthesis. Several mechanisms have been postulated to explain the phosphate regulation of antibiotic biosynthesis; the most important are schematically represented in Fig. 7.2:

a. Phosphate stimulates primary metabolism, thus channeling energy and substrates toward growth instead of synthesis of secondary metabolites.

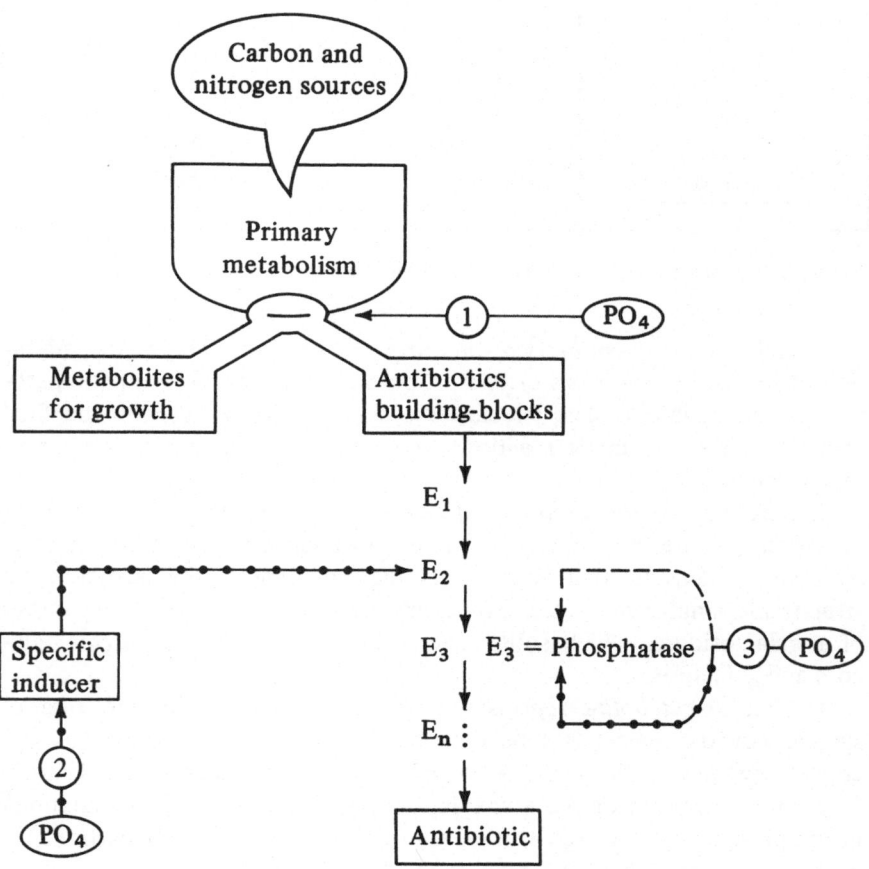

Fig. 7.2. Possible mechanisms of phosphate regulation of antibiotic biosynthesis. (•–•), Repression or induction of synthesis of enzyme; (– – –), control of activity of enzyme.

In this respect it is known that phosphate shifts carbohydrate catabolism from the hexose monophosphate pathway to the glycolytic pathway.
b. Phosphate controls the synthesis of a specific inducer of the synthesis of the antibiotic. This seems to be the case of factor A, which induces the synthesis of streptomycin and whose synthesis is controlled by inorganic phosphate.
c. Phosphate may inhibit the activity or repress the synthesis of phosphatases involved in the metabolism of phosphorylated precursor(s) of the antibiotic.

The phosphate regulation of antibiotic synthesis has been shown in different producing strains and with a variety of antibiotics. It is conceivable that different mechanisms may be operative in different organisms or under different physiological conditions and that more than one might be operative simultaneously. The alteration of one or more of the many control mechanisms is likely to be one of the most productive types of mutation, especially in the early stages of strain development.

f. Selection of Strains Resistant to the Antibiotic

Producing strains are often sensitive to their own antibiotic product. In some cases the mutants that are resistant to the toxic effects of the antibiotic are also high-producers.

g. Selection of Strains Resistant to Toxic Precursors

The classic case is that in which penicillin V is produced after addition of phenylacetic acid to culture broth. The precursor is toxic for the producing strain. The strains resistant to high concentrations of phenylacetic acid are often high-producers of penicillin V.

2. Genetic Engineering

The genetic information coding for the synthesis of an antibiotic may be "amplified" by increasing the number of one or more gene copies. This may result in cells with a higher rate of synthesis of the antibiotic. This is true only if the mechanisms controlling the expression of the genes do not limit the expression of the increased number of gene copies. Several techniques have been developed to achieved gene amplification. It is possible to "construct" (hence the term genetic engineering) microorganisms whose genome is constituted of traits coming from different strains: Such organisms are called *recombinants*. By crossing two different high-producing mutants (in which the ability to produce a high titer of antibiotic is due to different mutations), recombinants can be obtained that carry both parental strains.

It has been recently suggested that the synthesis of several antibiotics might be controlled by genes located in segments of DNA, separated from the chromosomes and called *plasmids*. Plasmids also can be moved from strain to strain.

A fascinating possibility of achieving gene amplification involves molecular cloning. This method consists of splicing a segment of DNA to a plasmid DNA that replicates and gives rise to several copies. In addition, by appropriately choosing the plasmids, the original piece of DNA can be transferred to different microorganisms such as *E. coli* whose physiology and genetics are well known, in which it may be expressed. The transfer of mammalian genes (such as those coding for insulin or dihydrofolic reductase) into *E. coli* and their expression have already been achieved.

However, it must be remembered that most, if not all, antibiotics are the products of several genes often dispersed throughout the chromosome. In this sense, an antibiotic is not the product of a gene but the product of the products (enzymes) of several or many genes. Therefore, it is unlikely that in the near future the transfer of the total information for the synthesis of an antibiotic from one species to another can be achieved easily.

B. Improvement of Fermentation Conditions

The rate of production of an antibiotic is also dependent on many nutritional factors: the nature and concentrations of the carbon and nitrogen sources, the concentrations of minor elements (phosphate, sulfate, etc.), the concentrations of many other substances, the pH of the medium, the oxygen tension, and amount of carbon dioxide dissolved in it. These factors can affect the production of antibiotic directly by interference with metabolic control mechanisms in their roles as precursors or effectors, in biosynthetic pathways, or indirectly by regulating the rate of cell growth. By manipulating systematically the components of the culture medium and the physical conditions of fermentation (aeration, shaking, temperature), it is possible to increase the productivity, that is, the production of antibiotic per unit of microbial mass per unit of time.

With larger scale production, however, the problem of scale-up arises. The studies involved in research and development of antibiotics are carried out with the instrumentation and apparatus present in the laboratory, in which producing organisms are grown in vessels containing from a few hundred milliliters to a liter. Industrial production, to be economically feasible, must take place in fermenters containing large volumes. There is the problem of converting the optimal fermentation process from the laboratory scale to the industrial scale. These studies are called "scaling-up" and are concerned with two different correlated aspects: (1) designing industrial fermenters that can reproduce the conditions in the laboratory giving maximal productivity; (2) modifications of the laboratory fermentation process to adapt it to the different, larger scale.

The need for and the difficulties involved in this last point derive from the fact that one cannot always increase the different parameters proportionally. For example, aeration and the energy needed for stirring vary proportionally with the volume of the fermenter, whereas dispersion of heat is proportional to the surface, which means that quite different times must be used for many operations of heating and cooling connected with sterilization. And the differences in the height of the liquid layer in the fermenter affect the times needed for exchange of carbon dioxide in different proportion. The development of the microorganism from the initial colony to a broth containing an equal number of cells per cubic centimeter requires not only more time, but a larger number of generations, and this may cause differences in productivity.

In recent years there has developed a branch of engineering known as bioengineering concerned with identifying the parameters important in the extrapolation from laboratory experiments to the commercial scale.

Chapter 8

The Use of Antibiotics

I. Chemotherapy of Infectious Diseases

The term chemotherapy refers to treatment of diseases of microbial origin by the systemic administration of antibiotics or other drugs. Basically, this involves inhibition of the multiplication of the infectious microorganism through selective toxicity, without interfering with the function of the host, thereby enabling the host's defense mechanisms to overcome the infection. It is beyond the scope of this book to undertake a detailed analysis of the methods or of the uses of chemotherapy, i.e., why a particular antibiotic is used for a given infection and how it is administered. However, it is important to analyze the basic principles of chemotherapy, which can be stated briefly as follows:

An antibiotic is therapeutically effective:

1. when it is used to treat or prevent infection caused by a susceptible organism,
2. when it can reach concentrations in tissue that sufficiently inhibit the growth of the infectious agent,
3. when the treatment is continued for a sufficient period of time,
4. when the antibiotic does not cause severe adverse reactions.

A. Principles of Chemotherapy

The various aspects to consider in selection of an antibiotic are outlined in Table 8.1.

Table 8.1. Aspects to Consider When Deciding on Antibiotic Treatment

Aspects	Considerations
Microbiological	Nature of infecting organism Its spectrum of sensitivity to antibiotics
Pharmacological	Dosage, interval between doses, duration of treatment
General clinical and toxicological	General condition of the patient: age, pregnancy, genetic factors, concomitant illness and treatment, liver and kidney function, condition of immune system, etc.
Adverse reactions	Suprainfections, aviaminosis, immune reactions, etc.
Epidemiological	Frequency of single and multiple resistants in environment in which patient lives Prevention of selection of resistant mutants and of spread of resistance

1. Microbiological Aspects

Obviously, antibiotics are used only when the disease to be treated is infectious (either established or strongly indicated to be so). It is also obvious that it is of no use to give antibacterial antibiotics for viral or fungal infections, and vice versa. No one antibiotic is active against all bacterial species, and, in fact, each antibiotic has its own specific spectrum of activity. In addition, within a single species of bacteria there are strains having different susceptibilities to the same antibacterial agent.

Some bacterial strains that are initially susceptible to a given antibiotic can become resistant through one or more of the mechanisms described in Chapter 4. Therefore, under ideal conditions, before beginning treatment with an antibiotic, one should know what organism is responsible for the infection and what its spectrum of susceptibility to different antibiotics is—its antibiogram (see Chapter 2).

However, the accurate analysis of antibiograms requires at least 2 days, and it may be difficult to interpret because contaminating nonpathogenic bacteria may also be present at the infection site. Therefore, in most cases it is necessary to begin treatment before the results of the antibiogram are available, choosing the antibiotic on the basis of the clinical characteristics of the infection, and selecting the proper one as soon as the results of the susceptibility determination are available.

It is important to remember that the infectious material that is submitted for the antibiogram analysis must be collected before beginning the antibiotic treatment.

In some cases it is easy to choose the antibiotic, as the clinical characteristics of the infection indicate rather definitely what the pathogenic agent may be and experience indicates the antibiotic of choice against that microorganism. This is the situation in the case of typhoid fever, for example, which is caused by *Salmonella* and may be treated with chloramphenicol, or in scarlet fever, caused by streptococci, which is treated with penicillin G or V. In other situations the choice is more complicated because the same type of illness may be caused by several infectious agents (e.g., septicemias and bacterial pneumonia) or by microorganisms that are susceptible to several antibiotics, so that the choice must be based on clinical and pharmacological considerations. Ideally, the antibiotic used should inhibit the growth of the infectious agent without interfering with the nonpathogenic bacterial flora, to avoid secondary effects. In other words, the antibiotic should be specific for the etiologic agent causing the infection. This is in contrast with the current tendency to prefer wide-spectrum antibiotics over those with narrower spectra. The preference for wide-spectrum antibiotics is well-founded only when one suspects or has shown that more than one bacterial species with different spectra of sensitivity (see section I.C.) are involved in the infection, or when the infectious agent has not been identified and one suspects that it may be Gram-negative.

Generally, it is not important whether the antibiotic is bactericidal or bacteriostatic, since its function is to block the spread of the bacteria and permit the body's defense mechanisms (cellular and immunological) to rid the body completely of the infectious agent. However, when the immune defenses are not very effective or in chronic conditions, a bactericidal antibiotic is preferable.

2. Toxicological Aspects and Secondary Effects

Although the microbiological aspects of the infection are of basic importance, they are not the only parameters the physician must take into consideration in choosing the antibiotic. There are numerous other aspects to be considered. In this chapter we discuss the possibility of adverse reactions that can be caused by antibiotic treatment.

We have stated many times that antibiotics are a widely diverse group of chemical substances, so that there is no toxic or adverse reaction that may be considered typical for the entire class. The contraindications for each family of antibiotics must be considered separate. The only secondary effects that might be considered to be group-specific are those that result from changes in the normal bacterial flora with which we live symbiotically. The wide-spectrum antibiotics, especially when given orally, may cause massive destruction of the intestinal bacterial flora, with one possible consequence being the abnormal proliferation of insensitive organisms, especially of fungi.

These situations are called suprainfections. Usually they are not very severe, but sometimes they are serious and require medical treatment. It should be noted that the same effect on the intestinal flora can occur whether a wide-spectrum antibiotic is given intramuscularly or intravenously if the antibiotic is excreted into the bile in an active form.

A less important negative effect of the alteration in the bacterial flora of the intestine is a possible vitamin deficiency, which is sometimes seen during prolonged treatment and which is due to destruction of the bacteria that synthesize vitamins. This inconvenience is easily overcome by vitamin administration.

These phenomena are not of course limited to the intestinal tract (there may be, e.g., a suprainfection of the vagina with trichomonads) and they are rarely seen when narrow-spectrum antibiotics are given, because these cause fewer changes in the normal bacterial flora.

The majority of antibiotics have very low toxicity. The cellular and molecular bases for this characteristic are obviously related to the selectivity of action, previously discussed with reference to the mechanism of action.

Some families of antibiotics, however, do have toxic effects or special side effects, such as the ototoxicity of the aminoglycoside antibiotics or the photosensitization of the tetracyclines. These aspects have been discussed separately in Chapter 5, within the description of the principal antibiotic families. Obviously, it is very important to take them into account in clinical practice as possible contraindications. In addition, all drugs, or all extraneous substances taken into the body, including antibiotics, can stimulate immune phenomena, with production of antibodies and hypersensitivity or allergic reactions. The frequency and the severity of these reactions vary greatly from one family of antibiotics to another and can also be quite different for different members of the same family. Hypersensitivity and allergic reactions may take different forms. The least important are usually skin manifestations (rashes, etc.), which usually do not necessitate stopping the treatment. Fever and leukopenia are more serious. The most serious are aplastic anemia, angioedema, and anaphylactic shock. An antibiotic or others of the same family should never be given to a patient who has already demonstrated hypersensitivity to it, but this is not a contraindication for giving antibiotics from a different family.

3. Pharmacokinetic Aspects

Antibiotics are effective only if they are present at the infection site in concentrations equal to or greater than the MIC. The dose and the frequency of treatment are recommended on the basis of careful pharmacological and clinical research and guarantee that this condition is fulfilled. If too low doses are given, or adequate doses are given at intervals that are too far apart, one may never achieve inhibitory concentrations or may achieve them for insufficient periods of time at the infection site and the treatment will be ineffective. In addition to this, doses of antibiotic that are too low and irregular treatments may select mutants with the multistep type of resistance, since this produces in the body a situation similar to that used in the laboratory for isolating mutants of that type by the method called "training" (see Chapter 4).

How long should treatment be continued? There are no general principles that can be applied to all cases. Only research and clinical experience can indicate the "safe" duration of treatment for each illness. Classic examples of improper duration of treatment often occur in pharyngitis. The patient appears with a red and sore throat, accompanied by a high fever. It is diagnosed as streptococcal pharyngitis and treated with penicillin. After 1 or 2 days of treatment, the clinical symptoms have disappeared and the treatment is stopped, usually on the patient's own initiative. Then, 2 or 3 days later, fever, redness, and pain return. The cycle of treatment, interruption, and recurrence continues for months, when treatment with the antibiotic for 10 days would have completely eliminated the infection. In addition, there is the serious risk of development of rheumatic endocarditis or glomerulonephritis. Generally speaking, recurrences should be attributed to treatment that was not sufficiently prolonged.

4. Epidemioecological Aspects: Resistance

Streptococcus pyogenes group A, *Streptococcus pneumoniae,* and *Treponema pallidum* have not yet, even after 30 years, developed any appreciable degree of resistance to penicillin.[1] Unlike these, increasing numbers of antibiotic-resistant strains of staphylococci have been isolated from human infections within short times after the introduction into clinical practice of every new antibiotic. The same is true for many pathogenic Gram-negative bacteria, both bacilli and cocci.

In infections with the latter microorganisms, the physician must keep up with the latest information about the distribution of resistance to the different antibiotics in the different bacteria for the geographical area in which he is working. Thus, a doctor who knows that 80% of the staphylococcal population in his region is resistant to penicillin will not give that antibiotic for treatment of staphylococcal infections, unless the patients isolate is first demonstrated to be a penicillin-susceptible strain. It should be noted that

1. The distribution of resistance to an antibiotic in a given bacterial population does not remain constant with time but varies as a function of the use of the antibiotic in that region.
2. The frequency of resistant strains increases with the selective pressure, i.e., when the use of the antibiotic increases.
3. The frequency of strains with transferable resistance is a function of

---

[1] Penicillin-resistant streptococci have recently been isolated in several countries.

the concentration of the microbial population. Because of this, the highest frequency of strains resistant to one or more antibiotics (see Chapter 4 on multiple resistance) is found in hospitals where these drugs are used the most and where there is the highest concentration of the microbial population.

B. Prophylaxis

Antibiotics are often used to prevent an infection before it occurs clinically. Although there is little doubt that preventing a disease is better than curing it, we must not forget that the indiscriminate use of antibiotics as prophylactic agents may cause more damage (because of selection of resistant strains, possible adverse effects, sensitization of patients) than benefits. Rational use of antibiotics in prophylaxis requires:

1. A condition in which infection can be expected to occur. Examples are: (1) extensive burn trauma (infection by *S. aureus* or *P. aeruginosa*); (2) surgical operations such as leg amputations (gas gangrene caused by *C. perfringens*), cardiac surgery (endocarditis from *S. aureus*), dental surgery in patients with prosthetic cardiac valve; (3) streptococcal infections of the throat, kidney infection, rheumatic fever; (4) healthy carriers of meningococci and *H. influenzae*. This is not an exhaustive list and other examples could be added.

2. Correct choice of the antibiotic to be administered. Since a well-defined infection is expected, it is not difficult to select the appropriate antibiotic on the basis of the sensitivity of the causative microorganisms. Both systemic and topical treatment should be considered when appropriate (such as after surgery or in burned patients).

3. Correct timing of the administration. It is obviously pointless to give a product with a short biological life, like penicillin, 2 days before a surgical operation. The ideal time of administration is that which ensures the presence of the antibiotic at the site of the possible infection when this may be initiated. This time can easily be predicted for surgery. In other cases, such as rheumatic fever, because of the impossibility of predicting the time of reinfection, the antibiotic coverage must be extended over a long period of time.

C. Antibiotic Combinations

The term antibiotic combination means the simultaneous administration of two or more antibiotics. The practice of combining several antibiotics or of combining antibiotics with other chemotherapeutic agents is as old as the antibiotics themselves.

1. Reasons for Antibiotic Combinations

There are many reasons (microbiological, clinical or the simplistic idea that two drugs are better than one) for many different combinations to be proposed in place of single treatments. Not all these reasons are acceptable. They can be classified as follows below.

a. Decreased Frequency of Appearance of Resistant Bacteria

The frequency of mutation toward resistance to two antibiotics with different mechanisms of action is the product of the frequency of resistance to each antibiotic. Thus, if the frequency of resistance to antibiotic A is, for example, 10^{-7} and that to antibiotic B is 10^{-8}, the frequency of appearance of resistance to both A and B at the same time is $10^{-7} \times 10^{-8} = 10^{-15}$. In the combined treatment with A and B, the organisms resistant to A will be eliminated by B and the resistant organisms to B by A. Only the resistant organisms to both A and B will not be inhibited by the combination, but these occur with an extremely low frequency, negligible for all practical purposes [in the example given, there would be one doubly resistant bacterium in a population of one million billion (10^{15}) bacteria].

b. Enlarged Antibacterial Spectrum of the Treatment

The reasoning followed here is more or less the following: If one combines two antibiotics with limited but different spectra of antibacterial activity, the sum of the two spectra will be a wider spectrum. There are essentially two clinical situations in which these types of combinations are necessary: Treatment of mixed infections and treatment of severe infections of unknown etiology.

c. Decreased Doses of Potentially Toxic Antibiotics

It is hoped that the combination of lower doses of each drug will have the same effectiveness as treatment with full doses of one, but with less toxicity.

d. Increased Therapeutic Effectiveness

Experience has shown that some but not all combinations of antibiotics have higher in vitro antibacterial activity and greater therapeutic effectiveness than treatment with the single components. Often, the combination has activity or effectiveness either equal to the sum of those of the individual components or lower. The bacteriological and molecular reasons for these different behaviors of combinations (synergism, additivity, antagonism) are discussed later.

2. Rational Approach to Designing Antibiotic Combinations

By far the great majority of the combinations used to date have been prepared empirically as new antibiotics became available. With the enormous amount of bacteriological, biomolecular, pharmacological, and clinical information available today about the antibiotics, it is possible to construct theories about what objectives might be obtained better with a combination and what mutual characteristics the components should have for the combination to achieve the objective.

a. Decreased Appearance of Resistant Organisms

It is essential that the bacterial pathogens one wants to control be sensitive to both the antibiotics in the combination. If not, it is better to use only that component to which the bacterium is sensitive. The cellular targets for each antibiotic should be different; otherwise, a mutation in the target giving resistance to one component will also make it resistant to the other. Therefore, one cannot use combinations of antibiotics that have the same mechanism and site of action and are cross-resistant.

Naturally, to be active, each antibiotic must be present in the infection site at a concentration equal to or higher than the MIC, and therefore full doses that guarantee levels higher than the MIC must be used unless the combination is synergistic. In addition, the two components must have compatible pharmacokinetics.

b. Widening the Antibacterial Spectrum

As we have said, this is the objective when one treats mixed infections, such as otitis media, peritonitis, and suppurative bronchiectasis. In this situation the two components must have different antibacterial spectra so that the sum of the two spectra is very wide. It could be that some of the bacterial species in the mixed population will be sensitive to only one of the components of the combination. It is necessary that the mechanism of resistance to the antibiotic in the insensitive strain is not the inactivation of the antibiotic.

Full doses, sufficient to guarantee levels of each antibiotic above the MIC, should be given. Obviously, the presence of one antibiotic should not increase the MIC for the other, which is to say that the two components are not antagonistic.

Today many wide-spectrum antibiotics are available (tetracyclines, penicillins and semisynthetic cephalosporins), and the failures of monotherapy with these are often due to problems associated with development of resistant organisms rather than to insensitivity of one member of a mixed population. Therefore, a rational combination aimed at decreasing the development of resistance can effectively resolve many cases with mixed infections.

The "widened-spectrum" combination is also used for treatment of infections of unknown etiology. It is to be repeated that each antibiotic treatment should be given only when there are definite bacteriological or clinical indications, to avoid useless treatments or even harmful results.

c. Decreased Doses of Components in the Mixture

It is important to note that the toxic effects of an antibiotic are often not dose-dependent and therefore, decreasing the dose often is not good practice. On the other hand, decreasing the dose may result in blood levels of antibiotic that are lower than the MIC, with consequent ineffectiveness and also with the possibility of selection of multistep resistants. Therefore, combinations designed for this goal are of doubtful usefulness.

d. Increased Therapeutic Effectiveness

When a homogeneous bacterial population is treated simultaneously with two antibiotics, there may be three types of result: synergism, additivity, or antagonism.

With *synergism*, the antibiotic effects of the combination are greater than the sum of the effects of each antibiotic alone. *Additivity* refers to an effect of the combination that is equal to the sum of the effects of the two. In the case of *antagonism*, one of the components decreases the effectiveness of the other (bacteriostatic or bactericidal). Although these three situations are easy to distinguish conceptually as separate entities, it is often complicated to measure them quantitatively. In addition, interference of one antibiotic with another may or may not be seen, depending on the parameter measured. For example, a compound can influence the bactericidal power of another without changing the MIC. Also, the nature of the interference is often dependent on the conditions under which it is determined; for example, the size of the bacterial inoculum.

In spite of these difficulties, one can often explain the biochemical reasons for these three different effects from knowledge of the mechanisms of action of the antibiotics. For example, the known antagonism between tetracycline and penicillin is due to the fact that penicillin is active only on growing bacteria. Tetracycline blocks growth, and the bacteria remain in a physiological phase insensitive to penicillin. The molecular explanation for synergism is usually less obvious. The most typical example of synergism is that of two chemotherapeutic agents with antimetabolite action, sulfamethoxazole and trimethoprim, in which the synergism is due to the so-called double metabolic block mentioned in Chapter 3. One can also have synergism when one antibiotic in the combination, at sublethal doses (those below the MIC), either directly or indirectly causes a change in the permeability to the other antibiotic, enabling it to reach inhibitory levels within the cell at a lower external concentration, thus lowering its MIC.

When there is synergy, the concentrations of each antibiotic in the combination needed to obtain inhibition of bacterial growth are only a fraction of the MIC for each antibiotic when used alone. Therefore, it is possible to treat an infection with a combination containing doses of each antibiotic lower than those needed in single treatment. It must be kept in mind that the pharmacokinetics of two components must be such as to insure the presence and the persistence in the infection site of those concentrations of each antibiotic at which synergism is seen.

Finally, it has been shown that combinations can be either more toxic or toxic in a different way than the components given singly. In addition, the presence of one drug can affect the pharmacokinetics of the other (absorption, distribution, metabolism, excretion). Therefore, every new combination must be considered to be a new drug and must be submitted to the studies of toxicology in animals that all drugs must undergo before being tried in man and then to studies of its pharmacokinetics and metabolism in man.

II. Uses Other than Human Pharmacology

In addition to their uses in human pharmacology, the antibiotics are widely used in animal husbandry and in veterinary medicine, and, in some countries, in agriculture. In these areas, the antibiotics are also used sometimes for purposes that are not strictly therapeutic and according to principles applied differently from when they are used in human medicine. In the next pages we briefly describe these different uses with their advantages, disadvantages, and potential risks. As in the preceding section, we do not give a detailed description of the uses of individual antibiotics in veterinary practice and in agriculture, for which we must turn to specialized publications on the subject, but limit ourselves to discussing the principles for their use.

A. Veterinary Use

1. Prophylaxis and Therapy

In veterinary medicine the antibiotics are widely used in prophylaxis and treatment of diseases in programs that are designed either for individual treatment or for mass treatment of the animals.

a. Individual Treatment

This type of intervention is practiced in treatment of small animals (dogs, cats, pets, etc.) and in treatment of large animals (cows, sheep, etc.) suffering from infections (pulmonitis, bronchitis, mastitis, etc.).

In treatment of small animals, the same criteria applied to chemotherapy of infections in humans are applicable and the same antibiotics are used.

In the treatment of large animals, the principles of human chemotherapy are widely applicable, but for animals designed for food production (meat, eggs) it is necessary to withdraw the antibiotic for a sufficient period of time before they are sent to slaughter or the eggs collected, so as to avoid ingestion of the antibiotics by the population using the commodity.

b. Mass Treatment

Prophylactic mass treatment may be given to all the animals in an establishment, whereas therapeutic mass treatment is given to all those present in a facility which has been affected by an infection. In this type of intervention, applied to the raising of different species of animals (chickens, pigs, calves), the antibiotics are given in the food or the drinking water (medicated supplements).

Mass treatment has the following advantages over individual treatment:

1. Simultaneous treatment of all the animals interrupts the infectious cycle and eliminates the foci of infection.
2. Treatment of animals with active infection and with asymptomatic or presymptomatic infections makes it possible to achieve cures without after effects that may reduce the economic value of the animals.
3. Treatment is facilitated because the antibiotics are given in the food or drinking water.

The disadvantages of mass treatment include the following:

1. It is difficult to obtain exact individual dosage.
2. There are difficulties related to the types of feed, the cycles, and the technology of breeding.

Even with these limitations, mass treatment is at present considered the most suitable and economic system for an effective attack on infectious diseases in intensive animal raising conditions.

2. Improvement of Production in Animal Husbandry

Antibiotics have been widely used for several years in animal nutrition to increase the rate of growth of the animals.

a. Growth Promotion

Addition of antibiotics to animal feed, common over the last 20 years, has no parallel in human medicine and should be discussed separately.

In 1948, it was first noticed that incorporation of small doses of antibiotics in the diet increased the weight gains in chickens. These experiments were confirmed innumerable times with a variety of antibiotics and different species of animals. As we have stated in many instances, the different antibiotic and chemotherapeutic compounds have nothing in common except their capacity to inhibit bacterial growth, so the effect of antibiotics on weight gain in animals must be mediated through their effects on the intestinal bacterial flora. This hypothesis was confirmed by the discovery that germ-free animals raised under sterile conditions do not show any response to antibiotic treatment and their rates of growth are equal to those of normal animals treated with antibiotics. When germ-free animals are artificially infected with bacteria present in the intestine of normal animals, their rate of growth decreases to the same level as that in the normal animals and can then be stimulated by incorporation of antibiotics in the diet. This effect on growth is probably due to one or more of the following causes:

1. inhibition of intestinal bacteria that produce toxins,
2. inhibition of bacteria that cause disease (asymptomatic),
3. inhibition of bacteria that destroy or sequester the protein or other essential nutrients in the diet,
4. stimulation, as a consequence of inhibition of part of the flora, of those bacteria that synthesize vitamins and other nutritional factors needed for growth of the host.

In the last analysis, the effect of the antibiotic consists of modifying the system constituted by the intestinal bacterial flora, with consequent alteration in the physiology of the animal. The final result is a greater rate of weight gain and a better conversion factor (less food consumed for an equivalent weight increase). This is translated into a consistent economic saving.

b. Hygienic and Sanitary Aspects of Antibiotic Use in Veterinary Medicine

It is not difficult to see that an intensive and massive use of antibiotics in animal husbandry could lead to conditions that are dangerous both for human and animal health. Figure 8.1 shows the most important and probable dangers connected with the use of antibiotics in animal husbandry. These dangers can be divided into two classes: bacteriological/epidemiological and pharmacological.

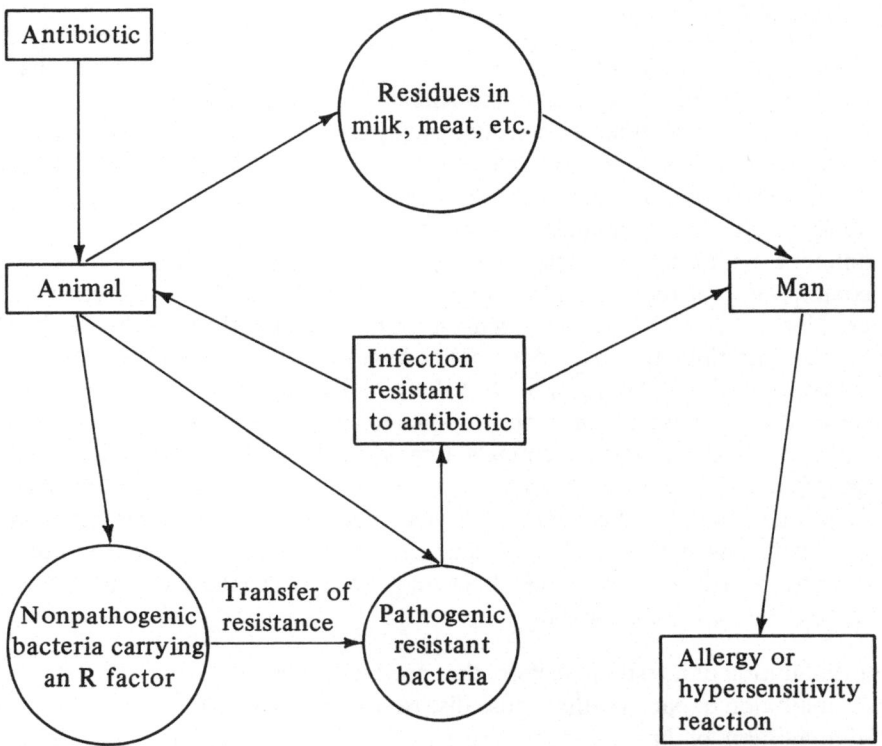

Figure 8.1. Dangers to human health connected with the use of antibiotics in animal husbandry.

b1. Bacteriological/Epidemiological   These consist of possible selection, direct or indirect, of germs pathogenic for humans or animals and resistant to antibiotics. Direct selection would occur when the antibiotic eliminated the sensitive strains and permitted the development and diffusion of the resistant ones. Indirect selection would occur when bacteria that are not themselves pathogenic for man, but are able to infect man, would transfer their own resistance carried by R factors (see Chapter 4) to pathogenic germs. This phenomenon has been repeatedly described for the Enterobacteriaceae.

b2. Pharmacological   With repeated and massive use of antibiotics in animal husbandry antibiotic residues can accumulate in the products of animal origin in amounts able to induce allergy or hypersensitivity in man. To avoid this, it is necessary to stop administering the antibiotic in the food for a sufficient period of time before the animal is sacrificed (withdrawal period). These periods are different for the different antibiotics and correspond to the time periods needed for the level of the antibiotic in the tissues to fall to concentrations which are considered to be safe.

With regard to the health of the animals, the most important danger of a pharmacological nature is that the antibiotic might interfere with the effects of other drugs.

c. Properties of the Ideal Growth-Promoting Antibiotic

In spite of the potential hazards for human and animal health, the advantages derived from use of antibiotics as growth factors in animal feeding are so great that to abandon them would have a disastrous effect. It is possible, however, that the pharmaceutical industry might discover and develop one or more antibiotics tailored for this particular usage that would minimize the dangers and be highly effective. Such antibiotics would have the following properties:

1. Microbiological properties.
   Spectrum: activity against those bacterial species in the intestine whose presence or prevalence has a negative effect on weight gain of the animal.
   Resistance: low emergence of resistant strains in species of coliform bacilli and salmonella; resistance preferably not associated with an R factor and, therefore, not easily transferred.
2. Pharmacokinetic properties. If possible, not absorbed and therefore limited to the intestinal tract. If absorbed, they should not concentrate in the tissues used as meat, in eggs or milk.
3. Ecological properties. They should be rapidly degraded to biologically inactive products after excretion.
4. Chemotherapeutic properties. They should not show cross-resistance with antibiotics used in human medicine.
5. Properties of usage. They should be compatible with the food; they should be stable when mixed in the food; they should not cause the food to be less palatable to the animal; they must be economical.

B. Agricultural Use

A certain number of antibiotics are used, especially in Japan, the U.S.A., and India, to control bacterial and fungal infections of plants of agricultural interest. The most commonly used antibiotics for this purpose are the following.

1. Streptomycin

This is used essentially against species belonging to the genera *Erwinia, Xanthomonas,* and *Pseudomonas* that cause infections in apples, pears, nuts, tomatoes, peppers, and beans in the U.S.A., and cotton, rice, and lemons in India. It is also effective for control of tobacco diseases caused by *Peronospora.*

2. Cycloheximide (or Actidione)

This is active only against fungi, not bacteria, and is used in control of fungal infections of cherries, roses and pine trees.

3. Blasticidin S and Kasugamycin

These are products developed in Japan exclusively for agricultural use, as substitutes for synthetic fungicides containing mercury. They are effective in control of infections of rice (rice blast) caused by *Piricularia oryzae*.

4. Aureofungin

This is a polyene antibiotic developed in India for agronomic use only. It is used in sterilization of seeds to prevent fungal diseases such as damping-off caused by *Pythium*.

III. Antibiotics as Research Tools

Many antibiotics, both those used in practice and those that have no therapeutic value, have been used and continue to be used as biochemical tools. There is an enormous literature on the subject, and we will limit ourselves to some examples. The usefulness of antibiotics as research tools lies mainly in their specific mechanisms of action.

1. Their ability to inhibit a definite informational molecule or enzyme make them very valuable for understanding the functions that this molecule has in the cellular metabolism. Many details of the protein synthesis mechanisms have been elucidated by using such inhibitors as chloramphenicol and puromycin. Differences in the mechanisms of protein synthesis in eukaryotic and prokaryotic cells have been revealed by the use of cycloheximide and chloramphenicol. The first indication of a difference between the RNA polymerases of eukaryotes and prokaryotes was obtained when the rifamycins were used. The degree of genetic autonomy of the mitochondria and chloroplasts was also made clear when the specific antibiotic sensitivities were observed.

2. One of the most complex tasks that geneticists started several years ago and that is still underway is to draw the genetic map of microorganisms, i.e., to establish the relative locations of the genes on DNA. Again antibiotics have proved to be invaluable tools. In fact, in all cases in which the resistance to an antibiotic is due to a modification of the target enzyme, locating the position of the "resistance character" in the genome is equivalent to locating the position of the enzyme. For example,

the position of the RNA polymerase gene on *E. coli* DNA was established by mapping the character resistance to rifampin in this microorganisms.

3. Several genetic operations make use of antibiotics, either as genetic markers (a definite pattern of resistance to some antibiotics makes a strain easily identifiable) or for selecting a microorganism with given characteristics. Penicillin, for example, is used in selection methods where the microorganisms able to grow under given conditions must be eliminated. Nongrowing cells are positively selected in this way.

Chapter 9

Antibiotics and Producer Organisms

What is the "natural" function of antibiotics, or, in other terms, what "evolutionary advantage" is conferred on a microbial strain by its ability to produce an antibiotic substance? Why is the ability to produce antibiotics more common in certain taxonomic groups than in others? During 30 years of intense antibiotic research several hypotheses have been put forth in answer to the above questions. In evaluating such hypotheses it is important to remember that antibiotics are a heterogeneous class of compounds in structure, biosynthetic origin, and mechanism of action and that they are the products of vastly differing organisms. Therefore, there may be more than one function of antibiotic substances, and they may confer different evolutionary advantages on different producing organisms.

Before analyzing the various hypotheses, a summary of the relationships between the systematic positions of the producing organisms and the nature of their antibiotics is necessary.

I. Classes of Antibiotics and Taxonomic Positions of Producing Organisms

It is important to attempt to establish a relationship between the taxonomic position of a producing strain and some properties of the antibiotic by-product, such as the chemical structure, the antimicrobial spectrum, or the mechanism of action. Several analyses have been made with this aim in mind. However, as more antibiotics are discovered, and as their characteristics become better defined, it becomes clearer that no simple relationship exists. However, there are several "general rules":

1. Most antibiotics are products of the secondary metabolism of three main groups of microorganisms: eubacteria, actinomycetes, and lower fungi. Only a few antibiotics are produced by higher fungi, algae, and plants and they generally show low activity and little specificity.
2. The actinomycetes produce the largest number and the greatest variety of the known antibiotics. More than 3000 different antibiotic substances have been isolated from this important group of microorganisms. The lower fungi produce several kinds of secondary metabolites and approximately 700 antimicrobial substances have been isolated from this group.

   The eubacteria, mainly spore-forming bacilli (*Bacillus*) and members of the genus *Pseudomonas,* produce a smaller number of antibiotic substances (400 have been described up to now) and these show little variety in their chemical structures, and most of the described bacterial antibiotics are polypeptides.
3. Frequently, a given strain produces a "family" of structurally and biosynthetically related substances (see Chapter 6). In addition, a given strain may produce two or more unrelated antibiotics.
4. The antibiotic production is not rigorously species-specific:

   (a) Different strains of the same species may produce completely different antibiotics. A classic example is *Streptomyces griseus.* Streptomycin (an aminoglycoside), novobiocin (a glycoside with a complex aromatic moiety), cycloheximide (aromatic structure derived from acetate), viridogrisein (a depsipeptide), griseoviridin (a lactone), candicidin (a polyene), and grisein (a sideromycin) are produced by different strains of this species.

   (b) On the other hand, the same antibiotic molecule can be produced by strains belonging to different taxonomical groups. For example, cycloserine has been isolated from both a *Streptomyces* and a *Pseudomonas* strain and penicillin N is produced by lower fungi (*Cephalosporium*) as well as by streptomycetes.
5. Nevertheless, the greater the taxonomic difference between two microorganisms, the lower is the probability that they will produce the same antibiotic molecule.
6. In addition to the above well-known rules, a relationship appears to exist between the taxonomic group of the producing organism and the biosynthetic pathway for its antibiotics. However, further research is needed to confirm this relationship. Some biosynthetic pathways of secondary metabolism occur generally (e.g., the capacity to activate and to condense amino acids to produce polypeptide antibiotics is found in eubacteria, actinomycetes, and lower fungi); others are present in only one of the three groups (e.g., practically all the known secondary metabolites originating from terpene synthesis are produced by fungi). Even within the actinomycetes there seems to be biosynthetic differentiation; for example, the biosynthesis of aminocyclitols, and therefore of aminocyclitol-containing antibiotics (aminoglycosides) is found much more frequently in the genera *Streptomyces* and *Micro-*

monospora than in other genera of the order Actinomycetales (*Nocardia, Actinoplanes,* etc.). It should be noted, however, that this relationship has only statistical and not absolute value. We hope that, with greater emphasis on research into the metabolism of antibiotic-producing organisms, it will become possible to establish a relationship between the biosynthesis of secondary metabolites (including antibiotics) and unique features of the organisms' metabolism.

For example, the ability (typical of fungi) to produce terpene antibiotics might be correlated with the presence of a metabolic pathway (the isoprene pathway) utilized by these organisms to produce the sterols of their cell membranes.

II. Paradox: How to Avoid Suicide

Many antibiotics are active against the organisms that produce them. In fact, if a suspension of spores or mycelium fragments is inoculated into a fresh medium containing the antibiotic that will eventually be produced by the organism, no growth occurs. We are dealing here with the apparent paradox of the ability of the strains to produce a large quantity of a substance that at low concentration inhibits its growth. This paradox is resolved by one or more of the following mechanisms: (1) repression of antibiotic synthesis during growth, (2) alteration of permeability, (3) inactivation of the antibiotic, (4) alteration of the intracellular target of the antibiotic.

1. Repression of Antibiotic Synthesis

Frequently, antibiotics are produced only after completion of the growth phase. This observation has led to the terminology *trophophase* and *idiophase.* The former refers to the period of vegetative mycelium growth, the latter to the period of antibiotic synthesis. The synthesis of antibiotics is often repressed by substances that favor rapid cellular growth, such as glucose (catabolite repression), ammonium ions (nitrogen repression), and phosphate. When the level of these nutrients is low, the rate of cell growth is slowed and antibiotic synthesis is derepressed. In this way, the antibiotic is synthesized only when the producing organism is "physiologically" insensitive to it.

2. Alteration of Permeability

It has been shown that certain high-producing strains differ from low-producing strains in their permeability to the antibiotic they produce. They are able to transport the antibiotic more efficiently and have lost the ability to assimilate or transport it from the medium.

3. Antibiotic Inactivation

Even in the presence of the above-mentioned mechanisms the cellular concentrations of the antibiotic can reach inhibitory levels. Suicide can be avoided by enzymatic inactivation of the antibiotic.

4. Alteration of the Target of the Antibiotic

In high-producing mutants, the cellular target of the antibiotic (ribosomes, RNA polymerase, etc.) is often structurally changed and, consequently, the producer strain is less sensitive to the antibiotic.

Frequently, more than one mechanism is operative in the same strain, as is the case for the high-producing mutants of rifamycin B. Rifamycin S, the central metabolite in rifamycin biosynthesis, which is a toxic compound, is inactivated by condensation with a glycol unit to produce rifamycin B, a nontoxic compound. Rifamycin B seems to be excreted more efficiently than rifamycin S. Finally, the RNA polymerase (the target of rifamycins) of high-producers is less sensitive to rifamycin S than is the enzyme of the wild-type strain.

III. Hypotheses about the Function(s) of Antibiotics in Producing Organisms

The major hypotheses put forth to explain the evolutionary significance of antibiotic production can be classified into three main groups: (1) elimination of metabolic waste products, (2) competition between organisms, and (3) regulation of metabolism.

1. Elimination of Metabolic Waste Products

According to this hypothesis, secondary metabolites are easily excretable compounds produced by enzymatic conversion of primary metabolites that accumulate in the cell under certain physiological conditions (e.g., the end of the growth phase). This hypothesis (which was proposed before the extensive regulatory mechanisms for controlling the cellular concentrations of metabolites were known) is no longer tenable, for the following reasons:

1. It does not explain why so many secondary metabolites have biological activity.
2. The quantity of antibiotic produced by wild-type microorganisms is generally very small and the need to eliminate small quantities of primary metabolites hardly justifies the existence of complex metabolic pathways.

3. The precursors of antibiotics are often substances with a central role in primary metabolism (e.g., acetate) that could be utilized more economically to synthesize storage products.

2. Competition Between Organisms

In the soil, where most antibiotic-producing organisms are found, life is competitive. The inhabitants must compete for the carbon, nitrogen, and phosphate (derived from degradation of plant material) necessary for their growth. Successful competition may be ensured either by metabolic specialization (development of enzyme systems utilizing special material sources) or by "inhibition" of the growth of other organisms through production and excretion of substances interfering with their metabolism (antibiotics). Unfortunately, very little is known about the synthesis of antibiotics in the natural environment. This hypothesis might be applicable to the actinomycetes which, as slow growers, must compete with faster growing microbes. It could explain why so many actinomycetes are able to produce antibiotics and why their antibiotics belong to so many different chemical and biological classes. The situation of the lower filamentous fungi is more complex. They are slow growers but, being physiologically different from bacteria, could compete with them by a less specific means, such as lowering the pH of the environment.

3. Metabolic Regulation

According to this hypothesis, antibiotics serve to interfere not with competitors' metabolism but with that of the producer. Their synthesis at specific moments in the development of the organisms would block specific metabolic functions and induce cellular differentiation.

This hypothesis seems tenable in the case of some peptide antibiotics produced by spore-forming bacteria. Here, antibiotic production occurs just prior to the onset of sporulation, and mutants unable to synthesize the antibiotic are generally also unable to sporulate. The antibiotic would play a role in effecting the transition from a vigorous vegetative cell to a dormant spore.

A metabolic function may be envisaged also for the ionophore antibiotics, such as the polyethers and the sideramines, which may play essential roles in scavenging from the environment and in transferring into the cell through chelation such ions as potassium and iron. However, at high concentrations they may become toxic to bacterial cells. The hypothesis of a regulatory role of antibiotics, although applicable in some cases, is unlikely to have general validity for the following reasons:

1. Many antibiotics are not active against the producing organisms (e.g., penicillins are produced by fungi and the natural compounds are active

only against bacteria; polyenes are produced by streptomycetes and active only against eukaryotes).

2. Although in some cases the association between the ability to produce antibiotics and to sporulate has been established, in others the two phenomena can be dissociated (i.e., mutants exist that have lost the ability to produce the antibiotic but retained sporulation). In addition, some antibiotic syntheses are not related to any specific phase of growth.

3. Some antibiotics have mechanisms of action that are difficult to reconcile with a regulatory role.

In conclusion, antibiotics are heterogeneous with respect to: chemical structure, taxonomic source, biosynthetic origin, mechanism of action, biological function to the producing organism, and mechanism by which microorganisms become resistant to them.

Further Readings

Chapter 1

Asselineau, J., Zalta, J. P. 1973. Les antibiotiques, structure et examples de mode d'action. Paris: Hermann.

Bérdy, J., Aszalos, A., Bostian, M., McNitt, K. L. 1980. Handbook of antibiotic compounds. Vols. I–IV. Boca Raton, Florida: CRC Press.

Korzibski, T., Koroszyk-Gindifer, Z., Kurilowicz, W. 1978. Antibiotics. Washington, D.C.: Annals of the Society of Microbiology.

Laskin, A. I., Lechevalier, H. A. 1973. Handbook of microbiology, Vol. 3. Boca Raton, Florida: CRC Press.

Reiner, R. 1974. Antibiotica und ausgewählte Chemotherapeutica. Stuttgart: Thieme Verlag.

Umezawa, H. 1967. Index of antibiotics from actinomyetes. Tokyo: University of Tokyo Press.

Waksman, A., Lechevalier, H. A. 1962. The actinomycetes, Vol. III. Baltimore: Williams and Wilkins.

Walter, A., Heilmever, F. 1969. Antibiotika-Fibel. Stuttgart: Thieme Verlag.

Zähner, H., Maas, W. K. 1972. Biology of antibiotics. Berlin: Heidelberg Science Library, Springer.

Scientific Journals

The Journal of Antibiotics, Japan Antibiotics Research Association, Tokyo.

Antimicrobial Agents and Chemotherapy, American Society for Microbiology, Washington, D.C.

The Journal of Antimicrobial Chemotherapy, British Society for Antimicrobial Chemotherapy, London.

Chapter 2

Balows, A. 1974. Current technique for antibiotic susceptibility testing. Springfield, Illinois: Thomas.

Berry, A. L. 1976. The antimicrobic susceptibility test. Principles and practices. Philadelphia: Lea and Febiger.

Lennette, E., Spaulding, E., Truant, J. (eds.). 1974. Manual of clinical microbiology, 2nd ed. Washington, D.C.: American Society for Microbiology.

Lorian, V. 1980. Antibiotics in laboratory medicine. Baltimore: Williams and Wilkins.

Reeves, D. S. (ed.). 1978. Laboratory methods in antimicrobial chemotherapy. London: Churchill Livingstone.

Chapter 3

Corcoran, J. W., Hahn, F. (eds.). 1975. Antibiotics, Vol. III. New York: Springer-Verlag.

Cozzarelli, N. 1977. The mechanism of action of inhibitors of DNA synthesis. *Annual Review of Biochemistry* **46**:641–668.

Franklin, J. J., Snow, G. A. 1975. Biochemistry of antimicrobial action. London: Chapman and Hall.

Gale, E. F., Clunliffe, E., Reynolds, P. E., Richmond, M. H., Waring, M. W. 1972. The molecular basis of antibiotic action. London: J. Wiley and Sons.

Gottlieb, D., Shaw, P. D. (eds.). 1967. Antibiotics, Vol. I. New York: Springer-Verlag.

Hahn, F. (ed.). 1979. Antibiotics, Vol. V. New York: Springer-Verlag.

Storm, D., Rosenthal, K., Swanson, P. 1977. Polymixin and related peptide antibiotics. *Annual Review of Biochemistry* **46**:723–763.

Tomasz, A. 1979. The mechanism of the irreversible effect of penicillins: How the β-lactam antibiotics kill and lyse bacteria. *Annual Review of Microbiology* **33**:113–138.

Westely, J. W. 1977. Polyether antibiotics. In: Perlman, D. (ed.), Advances in applied microbiology, Vol. 22. New York: Academic Press, pp. 177–220.

Chapter 4

Chopra, I., Howl, T. G. B. 1978. Bacterial resistance to the tetracyclines. *Microbiological Reviews* **42**:707–724.

Costerton, J. W., Cheng, K. J. 1975. The role of the bacterial cell envelope in antibiotic resistance. *Journal of Antimicrobial Chemotherapy* **1**:363–377.

Davies, J., Smith, D. 1978. Plasmid-determined resistance to antimicrobial agents. *Annual Review of Microbiology* **32**:469–518.

Falcow, S. 1975. Infectious multiple drug resistance. London: Pion Press.

Hahn, F. (ed.). 1976. Acquired resistance of microorganisms to chemotherapeutic drugs. Antibiotics and chemotherapy. Basel: S. Karger.

Kiser, J., Gale, G., Kemp, G. 1969. Resistance to antimicrobial agents. In:

Perlman, D. (ed.), Advances in applied microbiology, Vol. 11. New York: Academic Press, pp. 77–99.

Metsuhashi, S. (ed.). 1977. Factor R. Baltimore: University Park Press.

Metsuhashi, S. 1971. Transferable drug resistance factor R. Baltimore: University Park Press.

Chapter 5

Arcamone, A. 1978. Daunomycin and related antibiotics. In: Sammes, P. G. (ed.), Topics in antibiotics chemistry, Vol. 2. Chichester: Horwood, pp. 89–229.

Brufani, M. 1977. The ansamycins. In: Sammes, P. G. (ed.), Topics in antibiotic chemistry, Vol. 1. Chichester: Horwood, pp. 91–212.

Cooper, R. D. G. 1980. New β-lactam antibiotics. In: Sammes, P. G. (ed.), Topics in antibiotic chemistry, Vol. 3. Chichester: Horwood, pp. 39–203.

Cox, D., Richardson, K., Ross, B. 1977. The aminoglycosides. In: Sammes, P. G. (ed.), Topics in antibiotics chemistry, Vol. 1. Chichester: Horwood, pp. 1–90.

Daniels, P. 1978. Antibiotics (aminoglycosides). In: Encyclopedia of chemical technology, Vol. 2, 3rd ed. New York: J. Wiley and Sons.

Evans, R. M. 1965. The chemistry of antibiotics used in medicine. London: Pergamon Press.

Flynn, E. H. 1972. Cephalosporins and penicillins, chemistry and biology. New York: Academic Press.

Omura, S. et al. 1972. Relationships of structures and microbiological activity of the 16-membered antibiotics. *Journal of Medicinal Chemistry* **15:**1011–1015.

Perlman, D. (ed.). 1977. Structure–activity relationship among the semisynthetic antibiotics. New York: Academic Press.

Remers, W. 1980. The chemistry of antitumor antibiotics. Chichester: J. Wiley and Sons.

Umezawa, H. 1964. Recent advances in chemistry and biochemistry of antibiotics. Tokyo: Nissin Tosho Insatsu Company.

von Daehne, W., Gotfredsen, W. O., Rasnuskii, P. R. 1979. Structure–activity relationships in fusidic acid–type antibiotics. In: Perlman, D. (ed.), Advances in applied microbiology, Vol. 25. New York: Academic Press, pp. 95–144.

Chapter 6

Aberhart, J. 1977. Biosynthesis of β-lactam antibiotics. *Tetrahedron* **33:**1545–1559.

Corcoran, J. W. (ed.). 1980. Antibiotics, biosynthesis, Vol. IV. New York: Springer-Verlag.

Daum, S. J., Lemke, J. R. 1979. Mutational biosynthesis of new antibiotics. *Annual Review of Antibiotics* **33:**241–268.

Demain, A. L. 1974. Biochemistry of penicillin and cephalosporin fermentations. *Lloydia* **37**:147–167.
Gottlieb, G., Shaw, P. D. (eds.). 1965. Antibiotics, biosynthesis, Vol. II. New York: Springer-Verlag.
Martin, J. F. 1977. Biosynthesis of polyene macrolide antibiotics. *Annual Review of Microbiology* **31**:13–38.
Martin, F. J., Demain, A. 1980. Control of antibiotic biosynthesis. *Microbiological Reviews* **44**:230–251.
Omura, S., Nakagawa, A. 1975. Chemical and biological studies on 16-membered macrolide antibiotics. *Journal of Antibiotics* **27**:401–433.
Rinehart, K. L., Stroshane, R. M. 1976. Biosynthesis of aminocyclitol antibiotics. *Journal of Antibiotics* **29**:319–353.
Snell, J. F. 1966. Biosynthesis of antibiotics. New York: Academic Press.
Vaněk, Z., Hostaleck, Z. 1965. Biogenesis of antibiotic substances. New York: Academic Press.

Chapter 7

Demain, A. L. 1973. Mutation and the production of secondary metabolites. In: Perlman, D. (ed.), Advances in applied microbiology, Vol. 16. New York: Academic Press, pp. 177–202.
Spooner, D. F., Sykes, G. 1972. Laboratory assessment of antimicrobial activity. In: Methods in microbiology, Vol. 7.B. New York: Academic Press, pp. 211–276.
Waksman, S. A. 1969. Success and failure in the search for antibiotics. In: Perlman, D. (ed.), Advances in applied microbiology, Vol. 11. New York: Academic Press, pp. 1–16.
Weinstein, M., Wagman, G. (eds.). 1978. Antibiotics: Isolation, separation and purification. Amsterdam: Elsevier.

Chapter 8

Barker, B. M., Prescott, F. 1973. Antimicrobial agents in medicine. Oxford: Blackwell Scientific Publications.
Garrod, L. P., Lambert, P. H., O'Grady, F. 1981. Antibiotics and chemotherapy, 5th ed. Edinburgh: Churchill Livingstone.
Joint Committee on the Use of Antibiotics in Animal Husbandry and Veterinary Medicine. 1969. The Swann Report, Report Command n. 4190. London: Her Majesty's Stationery Office.
Kagan, B. (ed.). 1980. Antimicrobial therapy. Philadelphia: Saunders.
Kucers, A., Bennet, N. McK. 1979. The use of antibiotics, 3rd ed. London: Heinemann.
Mandel, A., Douglas, G., Bennett, J. (ed.). 1979. Principles and practices of infectious diseases. New York: John Wiley and Sons.
Woodbine, M. (ed.). 1977. Antibotics and antibiosis in agriculture. London: Butterworths.

Chapter 9

Gottlieb, D. 1976. The production and role of antibiotics in soil. *Journal of Antibiotics* **29**:987–1000.

Hopwood, D. A., Merrick, M. 1977. Genetics of antibiotic production. *Bacteriological Reviews* **41**:595–635.

Katz, E., Demain, A. L. 1977. The peptide antibiotics of *Bacillus:* Chemistry, biogenesis and possible functions. *Bacteriological Reviews* **41**:449–474.

Parenti, F., Coronelli, C. 1979. Actinoplanes and their antibiotics. *Annual Review of Microbiology* **33**:389–412.

Vining, L. 1979. Antibiotic tolerance in producing organisms. In: Perlman, D. (ed.), Advances in applied microbiology, Vol. 25. New York: Academic Press, pp. 147–165.

Wagman, G. H., Weistein, J. J. 1980. Antibiotics from micromonospora. *Annual Review of Microbiology* **34**:537–558.

Index

Springer Series in Microbiology

Editor: **Mortimer P. Starr,** Department of Bacteriology, University of California, Davis, California, U.S.A.

Thermophilic Microorganisms and Life at High Temperatures
T.D. Brock, University of Wisconsin, Madison
1978/xi, 465pp./195 illus./cloth
ISBN 0-387-**90309**-7

Bacterial Metabolism
G. Gottschalk, Universität Göttingen, Federal Republic of Germany
1979/xi, 281pp./161 illus./cloth
ISBN 0-387-**90308**-9

Ascomycete Systematics: The Luttrellian Concept
D.R. Reynolds (Ed.), Natural History Museum, Los Angeles
1981/vii, 242pp./122 illus./cloth
ISBN 0-387-**90488**-3

Bacterial and Bacteriophage Genetics: An Introduction
E.A. Birge, Arizona State University, Tempe
1981/xvi, 359pp./111 illus./cloth
ISBN 0-387-**90504**-9

General Nematology
A. Maggenti, University of California, Davis
1981/x, 372pp./135 illus./cloth
ISBN 0-387-**90588**-X

Basidium and Basidocarp: Evolution, Cytology, Function and Development
K. Wells and **E.K. Wells,** University of California, Davis
1982/xii, 187pp./117 illus./cloth
ISBN 0-387-**90631**-2